Springer Theses

Recognizing Outstanding Ph.D. Research

For further volumes:
http://www.springer.com/series/8790

Aims and Scope

The series "Springer Theses" brings together a selection of the very best Ph.D. theses from around the world and across the physical sciences. Nominated and endorsed by two recognized specialists, each published volume has been selected for its scientific excellence and the high impact of its contents for the pertinent field of research. For greater accessibility to non-specialists, the published versions include an extended introduction, as well as a foreword by the student's supervisor explaining the special relevance of the work for the field. As a whole, the series will provide a valuable resource both for newcomers to the research fields described, and for other scientists seeking detailed background information on special questions. Finally, it provides an accredited documentation of the valuable contributions made by today's younger generation of scientists.

Theses are accepted into the series by invited nomination only and must fulfill all of the following criteria

- They must be written in good English.
- The topic should fall within the confines of Chemistry, Physics and related interdisciplinary fields such as Materials, Nanoscience, Chemical Engineering, Complex Systems and Biophysics.
- The work reported in the thesis must represent a significant scientific advance.
- If the thesis includes previously published material, permission to reproduce this must be gained from the respective copyright holder.
- They must have been examined and passed during the 12 months prior to nomination.
- Each thesis should include a foreword by the supervisor outlining the significance of its content.
- The theses should have a clearly defined structure including an introduction accessible to scientists not expert in that particular field.

Michele Bianchi

Multiscale Fabrication of Functional Materials for Regenerative Medicine

Doctoral Thesis accepted by
University of Bologna, Italy

 Springer

Author
Dr. Michele Bianchi
National Research Council of Italy (CNR)
University of Bologna
Via Gobetti 101
40129 Bologna
Italy
e-mail: m.bianchi@bo.ismn.cnr.it

Supervisor
Prof. Dr. Fabio Biscarini
National Research Council of Italy (CNR)
University of Bologna
Via Gobetti 101
40129 Bologna
Italy
e-mail: f.biscarini@bo.ismn.cnr.it

ISSN 2190-5053
ISBN 978-3-642-27026-0
DOI 10.1007/978-3-642-22881-0
Springer Heidelberg Dordrecht London New York

e-ISSN 2190-5061
ISBN 978-3-642-22881-0 (eBook)

Cover design: eStudio Calamar, Berlin/Figueres

Printed on acid-free paper

Springer is part of Springer Science+Business Media (www.springer.com)

Parts of this thesis have been published in the following journal articles:

1. Valle F, Chelli B, Bianchi M, Greco P, Bystrenova E, Tonazzini I, Biscarini F (2010) Stable non-covalent large area patterning of inert teflon-AF surface: a new approach to multiscale cell guidance. Adv Eng Mater 12:B185–B191

2. Dionigi C, Bianchi M, D'Angelo P, Chelli B, Greco P, Shehu A, Tonazzini I, Lazar AN, Biscarini F (2010) Control of neuronal cell adhesion on single-walled carbon nanotube 3D patterns. J Mater Chem 4:2791–2800

3. Bianchi M, Limones Herrero D, Valle F, Greco P, Ingo GM, Kaciulis S, Biscarini F, Cavallini M (2011) One-step substrate nanofabrication and patterning of nanoparticles by lithographically controlled etching. Nanotechnology 22:355301

Part of Chap. 4 of this thesis is about to be published.

Supervisor's Foreword

Regenerative medicine needs new concepts and experimental tools to expand the knowledge about cell–cell and cell–environment interactions. In his thesis, Dr. Michele Bianchi showed that different kinds of signals (i.e. chemical, topographical, and electrical) can be spatially patterned with lengthscales ranging from several micrometers to few nanometers by using unconventional fabrication techniques. Further phenomena, like molecular diffusion, introduce a time scale in the chemical patterns, thus creating smooth chemical gradients that cells can follow in order to adhere to precisely defined regions of the substrate. For each signal, properly designed materials and fabrication methods have been chosen. The external signals have been found to be able to control cell adhesion and growth, opening the way for a systematic investigation of the environmental features affecting cell behaviour.

Bologna, August 2011

Prof. Dr. Fabio Biscarini

Preface

Regenerative medicine aims for a better understanding of the cause-and-effect relationship between cell behaviour and environmental signals. Environmental signals can be understood in terms of topographical, chemical and mechanical stimuli; electromagnetic fields; gradients of chemo-attractants; and haptotaxis. In order to mimic the natural cell environment in a tissue/organ, new concepts are needed. These new concepts must contribute to the rationalization of cell–cell and cell–surface contacts which are believed to represent key factors in cell behaviour. In order to develop new concepts, a spatial control of the structures featured in the environment is required. Nowadays unconventional micro- and nano-fabrication techniques allow the patterning of several biocompatible materials down to the level of a few nanometers. Patterning is not simply a deterministic confinement of a material; in many cases it allows a controlled fabrication of gradients of different nature. Indeed, gradients are emerging as one of the key factors guiding cell adhesion, proliferation, migration and even differentiation.

In this thesis I will describe a novel approach to multiscale patterning of bio-compatible functional materials in order to provide systems able to accurately control cell adhesion and proliferation. The behaviour of different neural cell lines in response to several stimuli, specifically chemical, topographical and electrical gradients will be presented. For each of the three kind of signals, I chose properly designed materials and fabrication and characterization techniques.

In Chap. 1 a brief introduction on the state of art of nanotechnologies, nano-fabrication techniques and regenerative medicine, with a particular focus on the thematic related to the regeneration of neural cells is shown. In Chap. 2 a detailed description of the main fabrication and characterization techniques employed in this work is reported. Chapter 3 (chemical control) describes an easy route to obtain accurate control over cell proliferation close to 100%. It is the first example of cell guidance on highly hydrophobic and chemical inert material such as Teflon-AF.

In Chap. 4 (topographical control) it is shown how the multiscale patterning of well-established biocompatible material as titanium dioxide provides a versatile and robust method to study the effect of local topography on cell adhesion and growth.

The third signal, viz. electric field, is investigated in Chap. 5 (electrical control), where the very early stages of neural cell adhesion are studied in the presence of modest steady electric fields.

In Chap. 6 (appendix) a new patterning technique, called Lithographically Controlled Etching (LCE), is proposed. I show how LCE can provide the micro/nanostructuring of a surface and its functionalization with nanosized objects at the same time. LCE has applications in the field of regenerative medicine thanks to the possibility to fabricate hierarchically structured materials as guides for cell growth. It is also useful for biosensing because of the possibility to fabricate metal and metal–oxide nanowire arrays, centimeters long with a lateral control of few tens of nanometers.

Bologna, August 2011   Dr. Michele Bianchi

Acknowledgments

First of all, I would like to express all my gratitude to my supervisor, Professor Fabio Biscarini, for having given me the possibility to enjoy his prestigious research group other than for his constant support and encouragements.

A special thanks to Dr. Francesco Valle, who demonstrated to be a topping scientific benchmark other than a faithful friend.

Many thanks to Dr. Massimiliano Cavallini for having financially supported my work in these years and for the large number of innovative and intriguing research ideas that he gave me. I would also like to thank Dr. Chiara Dionigi, for having taught me the importance of having the authorship of the ideas in the research world. Thanks to Arian, Stefano, Santiago, Cristiano, Crispin, Eva, Beatrice, Andreas, Dani, and all the other friends and people who succeeded in the group in these three years: it has been a pleasure to collaborate with them, other than having fun out of the lab. A special thanks to Carmen, the person that most of all shared with me the troubles and the satisfactions of these years. Last but not least I would like to thank my parents, my sister and my brother for having supported me in every decision and let me follow my way.

Finally, I want to dedicate this thesis to the chemist whose adventurous life instilled in me the flame of desire for scientific research and chemistry in particular, my late lamented grandfather Francesco Vescovi.

Contents

Chapter 1
Introduction

1.1 The Age of Nanotechnologies

Nanotechnologies have developed dramatically in the past decades, mainly due to huge improvements in fabrication and manipulation of objects with size of the order of billionths of a meter. The driving force behind nanotechnology, other than miniaturization, is the recognition that nanostructured material can have chemical and physical properties extremely different to those of bulk materials. The properties that we associate with bulk materials are averaged properties, such as the density and the elastic modulus in mechanics, the resistivity and the magnetization in electricity and magnetism and the dielectric constant in optics. At nanoscale dimensions the properties of materials no longer depend solely on composition and structure. Nanomaterials display new phenomena associated with quantized effects and with the preponderance of surfaces and interfaces. Quantized effects arise in the nanometric regime because the overall dimensions of objects are comparable to the characteriztic wavelength for fundamental excitations in materials. Such excitations include the wavelength of electrons, photons, phonons, and magnons; for example, electron wave functions in semiconductors are typically of the order of 10–100 nanometres. These excitations carry the quanta of energy through materials and thus determine the dynamics of their propagation and transformation from one form to another. When the size of structures is comparable to the quanta themselves, the effect of the excitations moving through and interacting with material structures is no longer negligible. This brings quantum mechanical selection rules into play, which are not applicable for structures of larger dimensions.

The term "nanotechnology" can be traced back to 1959, when the physicist Richard Feynman presented his visionary and prophetic talk entitled "There's Plenty of Room at the Bottom" at the annual conference of the American Physical Society. He speculated on the potential applications of nanosized materials, envisioning for example etching lines a few atoms wide with beams of electrons,

M. Bianchi, *Multiscale Fabrication of Functional Materials for Regenerative Medicine*, Springer Theses, DOI: 10.1007/978-3-642-22881-0_1,
© Springer-Verlag Berlin Heidelberg 2011

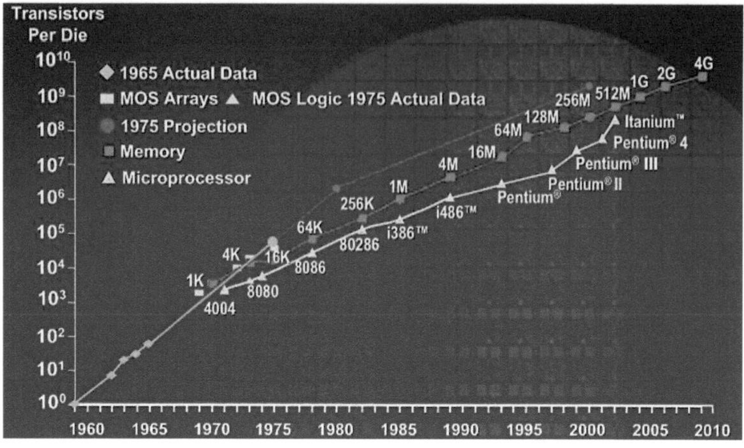

Fig. 1.1 Moore's law representation

effectively predicting the existence of electron-beam lithography, which is used today to make silicon chips. However, the very birth of nanotechnology as an applied discipline is represented by the invention of the Scanning Tunneling Microscope (STM) by Binnig and Rohrer in 1981 [1]. STM made it possible to observe and manipulate materials down to single atom level at the same time. In the beginning, the development of nanotechnologies was mainly driven by the microelectronics industry, which was focused on the miniaturization of integrated circuits. Following Moore's law (Fig. 1.1), which states that the density of transistors in a chip double approximately every eighteen months, electronic components and circuits have been downscaled to dimensions of hundreds of nanometers, thanks to the development of sophisticated techniques such as photolithography and electron beam lithography (predicted by Feynman). The strong competition between companies is directly linked to the minimum size of transistors. In other words, companies which pack more transistors onto a silicon chip and transfer this into computing power have the potential to realize sophisticated electronic systems. However, there are fabrication and theoretical limits to an infinite miniaturization process; scientists estimate that the mark would be around 10 nm.

1.2 Nanofabrication

Traditionally, material surfaces can be micro- and nanostructured following two opposite approaches: "top-down" and "bottom-up" (Fig. 1.2). The former plains to start from the bulk material and to progressively cut and shape it in order to obtain the final functional architecture. This approach is characteriztic of the conventional photolithographic techniques, which use the light to remove parts of the bulk material. Photolithography has been widely used in microelectronics since

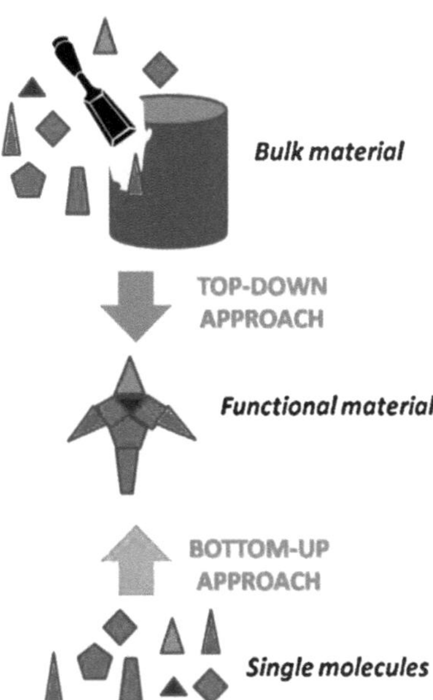

Fig. 1.2 Schematic representation of the two typical fabrication approaches

many years to realize microchips and hardware components. A predominant role in this category is played by UV-photolithography, that will be discussed more into detail in Sect. 2.1. However, the resolution of photolithography is limited by the wavelength of the light: that means that features smaller than few microns are not achievable with this technique. To overcome this limit, several alternative top-down techniques have been developed in the last decade, still exploiting light, as X-ray Lithography (XRL) [2] and Deep X-ray lithography (DXRL) [3], or based on De Broglie wavelength of the electrons as Electron Beam Lithography (EBL) [4]. All this methods allow to achieve very high resolutions (few tens of nanometer) but require expensive equipments; moreover few materials are suitable to be processed by these techniques.

The bottom-up approach is completely different: it starts directly from the molecules and is based on the their capability to self-assembly into regular structures under specific conditions. The driving force of this self-organization are specific and non-specific supramolecular and intermolecular interactions [5, 6]. These may vary from weak interactions (electrostatic and magnetic forces, hydrogen bonding) to strong interactions (covalent and ionic bonding). For example, non-covalent bonds are critical in maintaining the 3-D structure of large molecules, such as proteins, nucleic acids and artificial supramolecular structures, and are involved in many biological processes in which large molecules bind specifically but transiently to others. There are four commonly mentioned types of non-covalent interactions: hydrogen bonds, ionic bonds, van der Waals forces, and hydrophobic interactions. The energy released in the formation of non-covalent bonds is usually in the order of 1–5 kcal/mol. Dip-pen Lithography (DPL) [7], Local Oxidation Nanolithography (LON) [8], Nanoimprint Lithography (NIL) [9] and Soft-lithography [10] are examples of bottom-up techniques exploiting molecular self-organization. In particular, Soft-lithography will be discussed more into detail in Chap. 2, being the main fabrication technique I exploited in this thesis work. Thank to their flexibility and relative cheapness, bottom-up techniques have gained more and more appeal during these decades for many applications, i.e. electronics, optoelectronics, biotechnology and regenerative medicine.

1.3 Regenerative Medicine and CNS Diseases

Regenerative medicine is an emerging multidisciplinary field involving biologists, doctors, chemists and engineers that is likely to disrupt the ways we improve the health and quality of life for millions of people worldwide by restoring, maintaining, or enhancing tissue and organ functions [11]. In addition to therapeutic applications, such as controlled tissue growth inside the patient or outside and then transplanted, regenerative medicine has also many diagnostic applications, in which the tissue is cultured in vitro and used for example for testing drug metabolism and uptake, toxicity or pathogenicity.

The rationale of regenerative medicine and the key for its wide spreading in the next years, is the ability to fully exploit the skill of cells to build tissues. Interesting results have been shown at a preclinical level, but its massive implementation is still far. Indeed, most of the in vitro engineered tissues are still dysfunctional: this fact is mainly ascribed to the limited understanding of the fundamental principles underlying the complex cell-cell and cell-environment interactions, together with a restricted capacity of encoding bioactive signals in order to guide cells through the correct pathways of differentiation and biosynthesis. Numerous molecular signals affecting cellular behaviour were discovered in the last decades: accurate investigations of selected cues have highlighted the importance of chemical, topographical, electrical and more recently mechanical/viscoelastic features [12, 13]. Despite these efforts, little is known on the way cells display and exchange signals, in time and space, in order to elicit a specific cellular function.

Conventional signal screening devices are not suitable for a careful study of the effects of environmental signals on cell behaviour. This arises from their intrinsic incapability of producing precise signal gradients, low throughput screening and scalability. Nanotechnologies can provide a solution for these problems, thank to the wide submicrometric fabrication and characterization tool range. In particular, fabrication of gradients by means of effective and low cost techniques exploiting engineered nanodevices and nanostructures able to monitor, repair and control human biological systems at the molecular level, is strongly recommended [14].

A prominent topic in the field of regenerative medicine is the one concerning the study of the diseases of the central nervous system (CNS). Unfortunately the CNS can be subject to many diseases including infections as encephalitis and poliomyelitis, neurodegenerative diseases such as Alzheimer's disease and amyotrophic lateral sclerosis, autoimmune and inflammatory diseases such as

Fig. 1.3 Including the supplying of nutrients to the nervous tissue

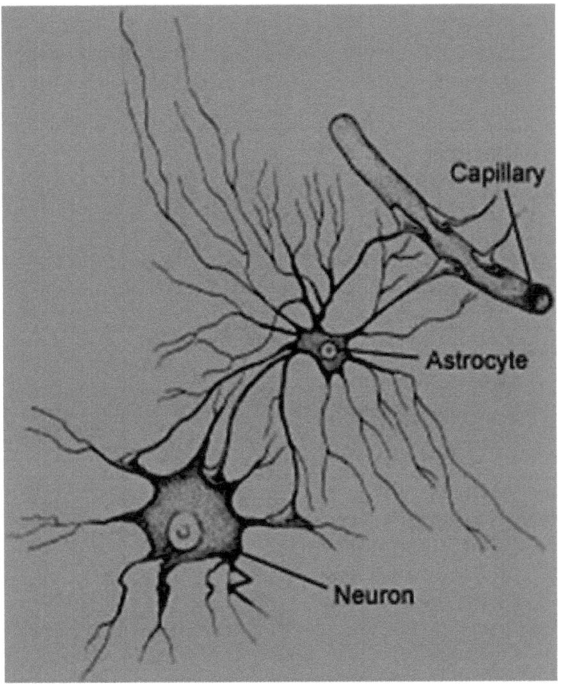

multiple sclerosis or acute disseminated encephalomyelitis and genetic disorders such as Krabbe's disease, Huntington's disease, or adrenoleukodystrophy. Lastly, cancers of the CNS can cause severe illness and, if malignant, have very high mortality rates.

Granted the importance of the issue, it is fundamental to find novel routes for investigating neural cell response with respect to the environmental signals, starting from a completely new approach. In this thesis, two human neural cell lines, viz. SH-SY5Y neuroblastoma and 1321N1 astrocytoma have been employed. They have been chosen because both these secondary cell lines have been extensively studied for many years as model of neural cell lines, more properly SH-SY5Y cells as neuronal lineage and 1321N1 cells as glial one; furthermore their cell culture conditions allow to obtain consistent adhesion and proliferation data [15].

The SH-SY5Y neuroblastoma cell line is a thrice-cloned sub-line of the bone marrow tumour biopsy-derived SK-N-SH cell line; neuroblastoma is a tumour derived from primitive cells of the sympathetic nervous system and is the most common solid tumour in childhood, with an annual incidence of about 650 new cases per year in the US [16]. The 1321N1 astrocytoma cell line is derived from the human brain astrocytoma, that is one of the most common gliomas, i.e. a malignant tumor affecting that particular kind of glia-cells called astrocytes [17]. In the past, the neuronal network was considered the only important part in the CNS, and astrocytes were looked upon as "gap fillers". More recently, the function of astrocytes has been reconsidered, and now they are thought to play a number of active roles in the brain, including the secretion or absorption of neuro-transmitters, biochemical support of endothelial cells forming the blood–brain barrier, thus supplying nutrients to the nervous tissue, preserving the extracellular ion balance, repairing the scarring process of the brain and spinal cord resulting from traumatic injuries (Fig. 1.3).

References

1. Binnig G, Rohrer H (1981) Helv Phys Acta 55:726–735
2. Smith HI, Schattenburg ML, Hector SD, Ferrera J, Moon EE, Yang IY, Burkhardt M (1996) Microelectron Eng 32:143–158
3. Ehrfeld W, Lehr H (1995) Radiat Phys chem 45:349–365
4. Liddle JA (2003) 739:19–30
5. Higgins AM, Jones RA (2000) Nature 404:476
6. Herminghaus S, Jacobs K, Mecke K, Bischof J, Fery A, Ibn-Elhaj M, Schlagowski S (1998) Science 282:916–919
7. Shim W, Braunschweig AB, Liao X, Chai JN, Lim JK, Zheng GF, Mirkin CA (2011) Nature 469:516–521
8. Garcia R, Martinez RV, Martinez J (2006) Chem Soc Rev 35:29–38
9. Chou SY, Krauss PR, Renstrom PJ (1996) J Vac Sci Technol B 14:4129–4133
10. Xia YN, Whitesides GM (1998) Angewandte Chem Int Ed 37:551–575
11. Reya T, Morrison SJ, Clarke MF, Weissman IL (2001) Nature 414:105–111
12. Roach P, Parker T, Gadegaard N, Alexander MR (2010) Surf Sci Rep 65:145–173
13. Discher DE, Janmey P, Wang YL (2005) Science 310:1139–1143
14. Keenan TM, Folch A (2008) Lab Chip 8:34–57
15. Jung DR, Kapur R, Adams T, Giuliano KA, Mrksich M, Craighead HG, Taylor DL (2001) Crit Rev Biotechnol 21:111–154
16. Brodeur GM (2003) Nat Rev Cancer 3:203–216
17. Giese A, Westphal M (1996) Neurosurgery 39:235–250

Chapter 2
Experimental Techniques

2.1 Fabrication Techniques

2.1.1 Photolithography

Photolithography basically consists in transferring geometric shapes from an optical mask to the surface of a silicon wafer by using light. To date, it is the main fabrication technique for the realization of integrated electronic circuits. The term Photolithography is a composed name that derives from the greek words "photo" (light), "lithos" (stone) and "graphein" (to write).

Its very first apparition can be dated to 1855, when Alphonse Louis Poitevin discovered that if a solution of potassium dichromate and albumin is let to dry on a lithographic stone and then exposed under a photographic negative, the parts exposed to light become insoluble and ink only adhere to those parts. Photolithography shares some fundamental principles with photography in that the pattern is made in the resist layer by exposing it to light, either directly or indirectly with a projected image using an optical mask. Subsequent stages in the process have more in common with etching processes than to lithographic printing.

2.1.1.1 The Photoresist

In the first step, a silicon wafer is chemically cleaned to remove particulate material on the surface as well as any traces of organic, ionic, and metallic impurities. Then a photoresist is applied to the surface of the wafer. High-speed centrifugal whirling of silicon wafers is the standard method for applying photoresist coatings and it is known as "Spin Coating". The parameters of the spin coater (revolutions per minute, acceleration, time) must be accurately set in order to get uniform thin films. There are two kinds of photoresist: positive and negative.

M. Bianchi, *Multiscale Fabrication of Functional Materials for Regenerative Medicine*, Springer Theses, DOI: 10.1007/978-3-642-22881-0_2,
© Springer-Verlag Berlin Heidelberg 2011

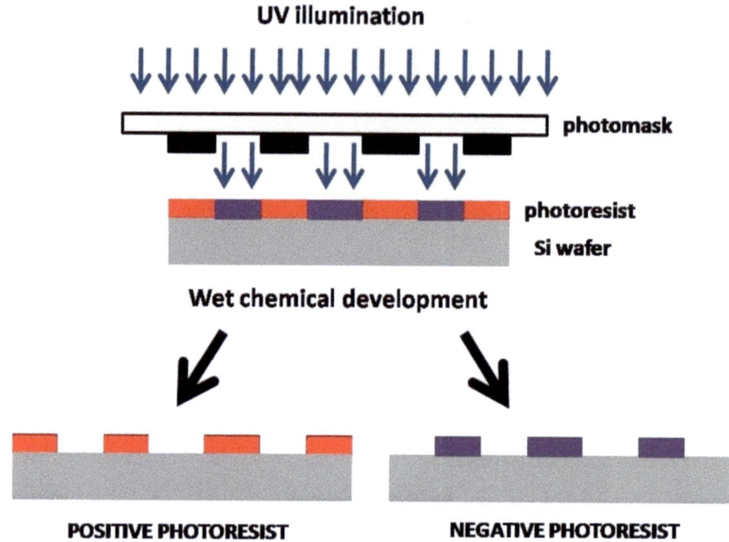

Fig. 2.1 Different pattern obtained starting from a positive or negative photoresist layer

For positive resists, the resist is exposed to UV light wherever the underlying material is to be removed. Hence the UV-exposure changes the chemical structure of the resist so that it becomes more soluble in the developer. The exposed resist is then washed away by the developer solution, leaving windows of the bare underlying material. In other words, "whatever shows, goes", with the mask containing an exact copy of the desired pattern. Negative resists behave in just the opposite manner. Exposure to the UV light causes the negative resist to undergo polymerization reactions thus becoming more insoluble. Therefore, the negative resist remains on the surface wherever it is exposed, and the developer solution removes only the unexposed portions. Masks used for negative photoresists, therefore, contain the inverse (or photographic "negative") of the pattern to be transferred. Figure 2.1 shows the different patterns generated with a negative and positive resist.

2.1.1.2 Pre-Baking

Pre-baking (or soft-baking) is a critical step, after the photoresist spin coating and before the UV illumination, during which almost all of the solvents are removed from the photoresist according to a precise heating protocol. Pre-baking plays a very critical role in photo-imaging because the photoresist coating become photosensitive only after this step. Overpre-baking will degrade the photosensitivity of resists by either reducing the developer solubility or actually destroying a portion of the sensitizer. Underpre-baking will prevent light from reaching

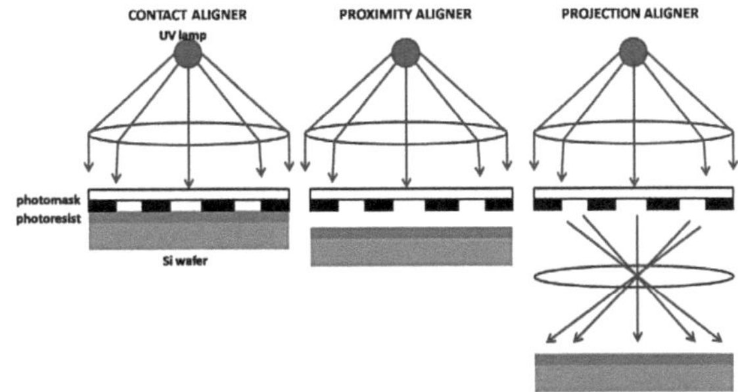

Fig. 2.2 Principal mask alignment methods

the sensitizer. Positive resists are incompletely exposed if considerable solvent remains in the coating. This underpre-baked positive resists is then readily attacked by the developer in both exposed and unexposed areas, causing less etching resistance.

2.1.1.3 Mask Alignment

Another basic step in the process is the mask alignment. There are three primary exposure methods: contact, proximity, and projection. They are schematically represented in Fig. 2.2.

In contact printing, the resist-coated silicon wafer is brought into physical contact with the glass photomask. The wafer is held on a vacuum chuck, and the whole assembly rises until the wafer and mask contact each other. The photoresist is exposed to UV light while the wafer is in contact position with the mask. Because of the contact between the resist and mask, a very high resolution can be achieved (e.g. 1 μm features in 0.5 μm of positive resist). One drawback is that debris, trapped between the resist and the mask, can damage the mask and cause defects in the pattern. The proximity exposure method is similar to contact printing except that a small gap (10–25 μm) is maintained between the wafer and the mask during exposure to minimize mask damages. Approximately 2–4 μm resolution is achievable with proximity printing. Projection printing instead avoids mask damage entirely. An image of the patterns on the mask is projected onto the resist-coated wafer, which is many centimeters away. In order to achieve high resolution, only a small portion of the mask is imaged. This small image field is scanned or stepped over the surface of the wafer. Projection printers that step the mask image over the wafer surface are called step-and-repeat systems (1 μm max resolution).

2.1.1.4 Development and Post-Baking

The goodness of the development process strongly depends on the initial exposure conditions (time, beam energy) and photoresist characterizctics (thickness, pre-baking protocol). At low-exposure energies, the negative resist remains completely soluble in the developer solution. As the exposure is increased above a certain energy, most of the resist film remains after development. At exposures two or three times the threshold energy, very little of the resist film is dissolved. For positive resists, the resist solubility in its developer is finite even at zero-exposure energy. The solubility gradually increases until, at some threshold, it becomes completely soluble. These curves are affected by all the resist processing variables: initial layer thickness, prebake conditions, developer chemistry, developing time, and others. Post-baking (or hard-baking) is the final step in the photolithographic process. This heating step is often necessary in order to harden the photoresist and improve its adhesion to the wafer surface.

2.1.2 Soft-Lithography

Although Photolithography is today the dominant technology for microfabrication, it has several drawbacks: it is not an inexpensive technology; it is poorly suited for patterning non-planar and three-dimensional structures; it provides little control over the chemistry of the surface thus being not versatile in generating patterns of specific chemical functionalities on surfaces; finally it is directly applicable to a limited set of photosensitive materials. During the 1990s George M. Whitesides and co-workers investigated and developed a new approach for rapid, low-cost and versatile micro-nanofabrication that they named Soft-lithography [1]. The break-through idea was that micro/nanopatterning relies on the interplay between capillary forces acting between an elastomeric stamp and a substrate and the self-organization of the molecules on the substrate that is mainly governed by non-covalent interactions (Fig. 2.3).

Nowadays Soft-lithography allows patterning of a variety of molecules and materials, including adhesion proteins [2], nucleic acids [3], ceramic [4] and polymeric materials [5]. Soft-lithography encompass several techniques sharing as key element an elastomeric microfluidic as stamp. The principal ones are "Microcontact Printing" (μCP) [6], "Replica Molding" (REM) [7], "Micro-transfer Molding" (μTM) [8], "Micromolding in Capillaries" (MIMIC) [9] and "Solvent assisted Microconctact Molding" (SAMIM) [10]. The common feature is the presence of an elastomeric material as stamp or mould. The principal elasto-meric material is polydimethylsiloxane (PDMS, Fig. 2.4), but polyurethanes, polyimides, and phenolic resins have been used [11].

Fabrication of channels in PDMS is particularly straightforward since it can be cast against a suitable mold with sub-micrometer fidelity [12]. PDMS is even more than a structural material: its chemical and physical properties make possible fabrication of devices with useful functionality (Table 2.1).

In particular its very low surface free energy allows the PDMS to percolate even into very small recesses enabling the reproduction of nanometer topographical details. Hence the resolution limit of Soft-lithography is set only by the materials properties (e.g. Van der Waals interactions, solubility, diffusivity) rather then by sample transparency and optical diffraction as for conventional Photolithography.

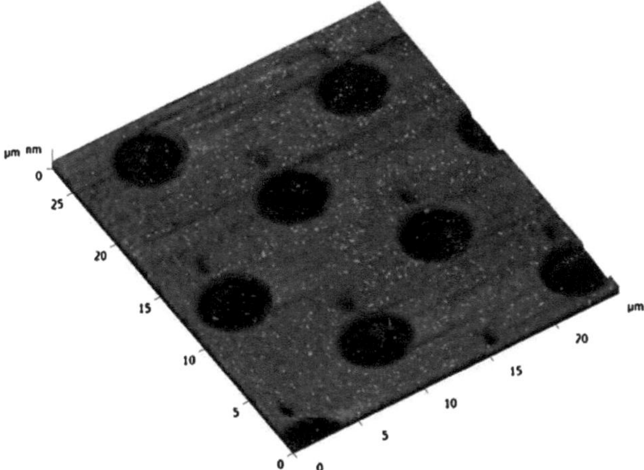

Fig. 2.3 Pattern of micrometric holes on silicon oxide surface obtained by Soft-lithography (AFM image)

Fig. 2.4 Siloxane backbone of PDMS

In Fig. 2.5 the entire cycle of soft-lithographic fabrication is reported.

Starting from the original idea, the pattern is transferred to a CAD file and printed on a transparent sheet of polymer with a commercial image setter. This patterned sheet is used as a mask to prepare the master in a thin film of photoresist; a negative replica of this master with an elastomeric material becomes the stamp or mold for Soft-lithography. One can start also from a commercial mask (usually of Chrome/quartz or polyester). The overall cycle from design to stamp takes less than 24 h to be complete. However, once one has the master, tens of stamp can be obtained by REM before master deteriorates, allowing a great time-saving. This is another great advantage of Soft-lithography over conventional Photolithography.

2.1.3 Replica Molding

"Replica molding" (REM) allows a faithful duplicate of the topographic information present in a master, even if three-dimensional. REM with an appropriate material such as PDMS enables replicating highly complex master structures with nanometric features in a simple, reliable and inexpensive way. In fact the fidelity of this process is largely determined by Van der Waals interactions, wetting, and

Table 2.1 Principal physicochemical properties of PDMS

Property	Charactcristic
Optical	Transparent; UV cutoff, 240 nm
Electrical	Insulating; breakdown voltage, 2×10^7 V/m
Mechanical	Elastomeric; tunable Young's modulus, typical value of ~ 750 kPa
Thermal	Insulating; thermal conductivity, 0.2 W/(m K); coefficient of thermal expansion, 310 μm/(m°C)
Interfacial	Low surface free energy ~ 20 erg/cm^2
Permeability	Impermeable to liquid water, permeable to gases and nonpolar organic solvents
Reactivity	Insert; can be oxidized by exposure to a plasma; $Bu_4N + F$-((TBA)F)
Toxicity	Nontoxic

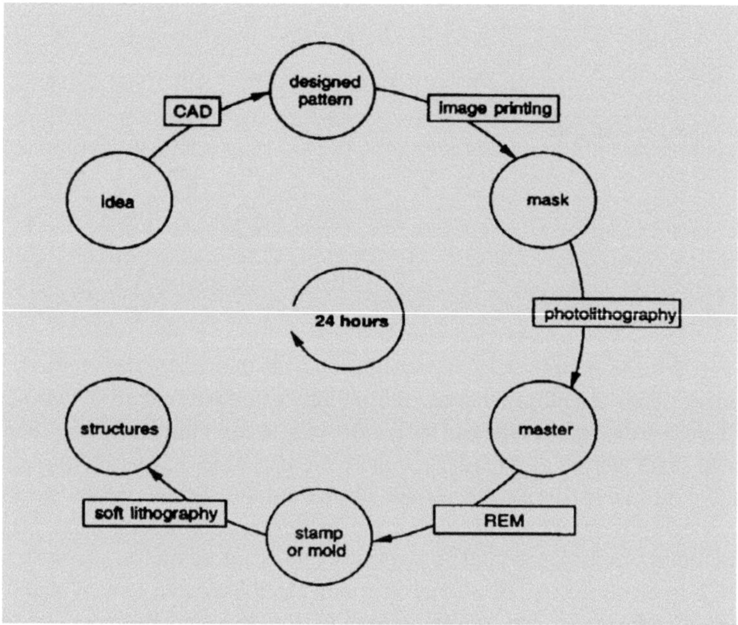

Fig. 2.5 The fast prototyping procedure for Soft-lithography

capillary forces that are short range interactions and allow more accurate replication of small features (<100 nm) than Photolithography. Thermally or UV-curable pre-polymers, have a shrinkage of less than 3% on curing as they usually do not contain solvent; the cured polymers, therefore, possess almost the same dimensions and topologies as the channels in the PDMS mold.

In Fig. 2.6 the general replication procedure is reported. The molds are prepared by casting against rigid masters a mixture of PDMS with a curing agent (generally in a ratio 10:1). The PDMS/master is put in a oven at 90°C for 3 h to

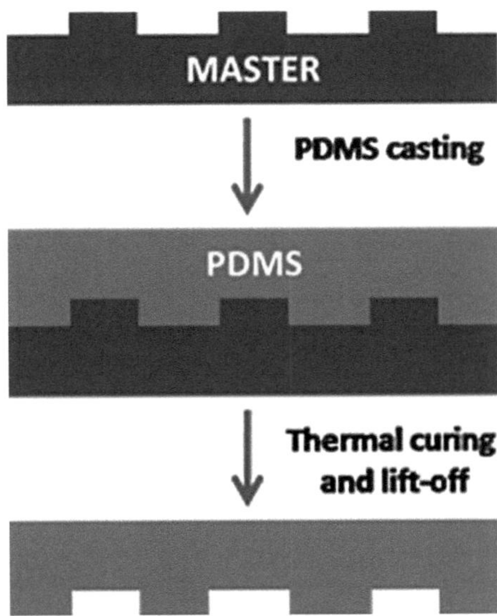

Fig. 2.6 Replica molding of the master with PDMS

complete the curing process. Then the mold is peeled off and it is ready to be use as stamp. This process can be replicated tens of time on the same master without introduction of defects.

2.1.4 MicroMolding in Capillaries

The general operation of "MicroMolding in Capillaries" (MIMIC) is reported in Fig. 2.7.

A patterned PDMS presenting open-end microchannels is brought into conformal contact with the substrate (Fig. 2.7a). A low viscosity solution is supplied to the entrance of the capillaries at one side of the mould (Fig. 2.7b).

The solution is drawn into the microchannels by Laplace pressure (Fig. 2.7c). In principle, no residue layer can form in areas where adhesive conformal contact between mould and substrate is already established. After solvent evaporation, the mould is removed to reveal patterned microstructures of the solute (Fig. 2.7d). Interestingly, capillaries with closed ends can also fill completely if they are short: the gas in them appears to escape by diffusing into the PDMS. MIMIC is applicable to patterning a broader range of materials than is Photolithography, including UV-curable (or thermally curable) prepolymers that have no solvents [13], precursor polymers to glassy carbon [14], sol–gel materials [15], inorganic salts [9], polymer beads [13] and biologically functional macromolecules [16].

For a tubular channel with a hydraulic radius R, the rate of penetration dz/dt is expressed by the Washburn equation [17]:

Fig. 2.7 Schematic representation of the MIMIC process

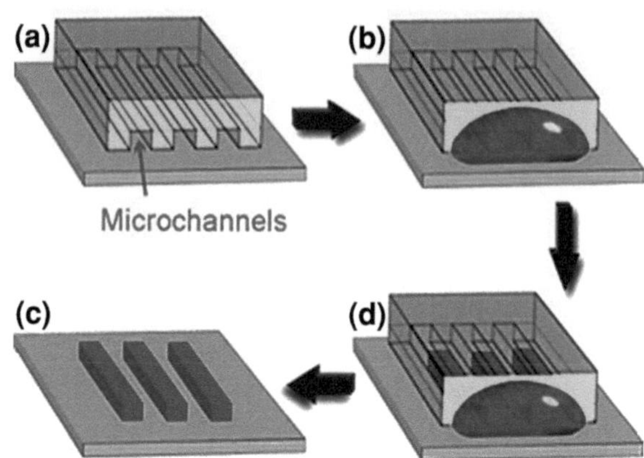

$$\frac{dz}{dt} = \frac{R\gamma_{LV}\cos\theta}{4\eta z} = \frac{R(\gamma_{SV} - \gamma_{SL})}{4\eta z}$$

Here z is the length of the liquid capillary inside the channel, η is the viscosity of the penetrating liquid, R is the ratio between capillary volume and the surface area of the channel, and θ is the contact angle of the fluid meniscus inside the capillary. The surface tensions γ_{LV}, γ_{SV} and γ_{SL} are the surface tensions between liquid and air, channel wall and air, and channel wall and liquid, respectively. It follows after integration that:

$$z(t) = \sqrt{\frac{R(\gamma_{SV} - \gamma_{SL})t}{2\eta}}$$

The equation shows that although the capillary force of a channel increases with decreasing hydraulic radius R, this effect is more than counterbalanced by the increased friction exerted by the channel walls, so filling rates are lower in smaller channels. A generally applicable method to increase the filling rate is to reduce the viscosity of the precursor solution by increasing the temperature during the MIMIC process. The main driving force for MIMIC is the free energy change ΔG upon filling the channel with a fluid. For a square channel with width and height a, the free energy gain can be approximated by:

$$\Delta G(t) = -az(t)\gamma_{LV}(3\cos\theta_{mould} + \cos\theta_{substrate})$$

where θ_{mould} and $\theta_{substrate}$ are the contact angles of the liquid with the surfaces of the mould and the substrate, respectively.

The shape of the imbibition front of liquid precursors has been studied in detail [18]. Depending on the surface energy of the channel wall γ_{SV}, different spreading regimes can be observed, as illustrated in Fig. 2.8.

Liquids penetrating inside a channel with low surface energy walls, show capillary fronts that advance as a whole. But as the surface energy of the channel

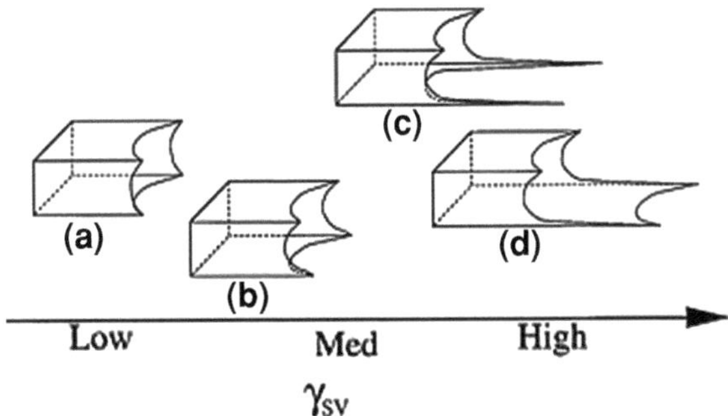

Fig. 2.8 Schematic representation of different spreading regimes observed in MIMIC. Shapes of the penetrating liquids in PDMS capillaries are formed inside **a** walls with low γ_{SV} **b** and **c** walls with medium γ_{SV}, and **d** walls with high γ_{SV}

increases, solute structures advancing in front of the macroscopic body of liquid are observed, especially in the corners between mould and substrate. Some of these structures include slipping films and shoulders. Similar regimes have been observed with differences in the velocity of imbibition on surfaces of constant γ_{SV}.

2.1.5 Lithographically Controlled Wetting

"Lithographically Controlled Wetting" (LCW) has been developed by researchers of ISMN-CNR of Bologna [19]. LCW is a micro/nanofabrication process particularly attractive for soluble materials because it allows one to exploit self-organization of soluble functional materials with the spatial control provided by the stamp features. The stamp can be made of either rigid materials, such as metals, ceramics or rigid polymers, or by soft materials, such as polycarbonates or PDMS. LCW is suitable for large-area nanopatterning and is sustainable because of its simplicity and high transfer rate. LCW represents a valid alternative and often a finer route to obtain nanosized structures with respect to MIMIC: in fact, while the latter is based on capillary forces that under specific surface tension conditions do not allow the filling of extremely small channels, the former is based on wetting-dewetting processes. Dewetting is the opposite of the process of the spreading of a liquid on a substrate (wetting): it describes the rupture of a thin liquid film on the substrate (either a liquid itself, or a solid) and the formation of a discontinuous film made of droplets. The factor determining the wetting and dewetting is the so-called spreading coefficient S, which is defined as:

$$S = \gamma_{SO} - \gamma_{SL} - \gamma$$

where γ_{SO} is the surface tension substrate/air, γ_{SL} is the surface tension film/substrate and γ is the surface tension film/air. Dewetting is driven by several mechanisms such as nucleation and growth of holes [20], ripening [21], spinodal dewetting or combination of them [22]. Although a spatial correlation between the features generated by dewetting usually emerges by advanced image analysis

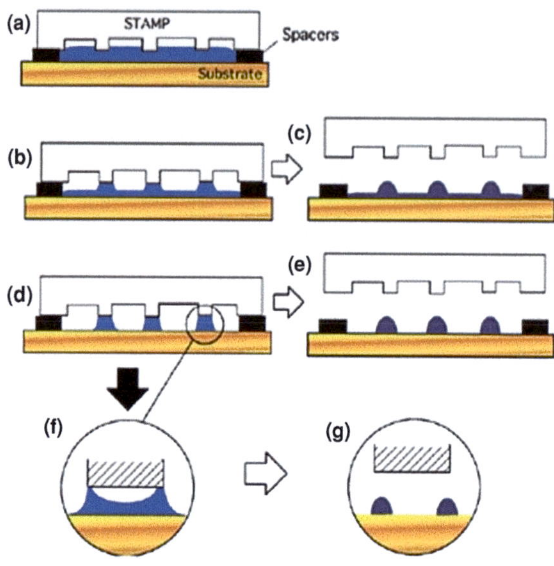

Fig. 2.9 LCW basic steps. Depending of experimental conditions, different patterns can be obtained starting from the same stamp. Spacers can be not present

(e.g. measuring the radial distribution of its height–height correlation function [23], dewetting can generate an ordered pattern of droplets or stripes [24].

LCW general operation is represented in Fig. 2.9.

When a stamp presenting topographical features is placed in contact with a liquid thin film spread onto a substrate (Fig. 2.9a), the fluid layer develops instabilities, where the capillary forces pin the solution to the stamp protrusions, giving rise to an array of menisci (Fig. 2.9b). As the critical concentration is reached, the solute precipitates from the solution onto the substrate within the menisci, giving rise to a structured thin film that replicates the protrusions of the stamp (Fig. 2.9c, e). In particular low concentration regimes (Fig. 2.9d), the menisci can be splitted under the same protrusion thus obtaining smaller structures (Fig. 2.9f, g).

The interplay of protrusion size with the physical length scales of self-organization and growth phenomena (e.g., correlation length) gives rise to the deposition of a material into domains whose lateral dimension ranges from micrometers to nanometers. Since the concentrations of the solutions used in LCW are usually low (thus the quantity of deposited material is small), the nanostructures often end up with a thickness equivalent to a few molecular layers. In this regime, growth phenomena are less prone to develop defects due to roughening, secondary nucleation, and orientational transitions. Thus, LCW allows to deposit highly ordered low-dimensional small-size domains, which can be used to tailor the properties of the material by size [25].

Since the described process does not rely on specific interactions between the molecules and the surface, LCW is of general application in a large variety of soluble materials such as rotaxanes [26], discotic liquid crystals [27], molecular magnets [28], and organic semiconductors [29]. Using suitable stamps, this technique can be used also with aggressive chlorinated and fluorinated solvents.

2.2 Characterization Techniques

2.2.1 Atomic Force Microscopy

2.2.1.1 Working Principle

In 1982 G. Binning, H. Rohrer, C. Gerber and E. Wiebel developed the scanning tunneling microscope, opening a new era in the surface science field. The limitation of this technique is the possibility to scan only conductive surfaces. The introduction of the Atomic Force Microscope (AFM), based on the atomic interactions rather than on the tunneling current, overcame this problem allowing the investigation of almost all the materials with physically relevant surface morphology. In more recent years other scanning probe techniques directly related to AFM have been introduced to probe also electrical, magnetical and elastic properties.

AFM is a relatively compact instrument, especially if compared to electron microscopies. It does not require vacuum to operate and can be placed on any stable workbench in a laboratory.

The key elements of an AFM are: (1) a local probe interacting with the surface; (2) a piezoelectric actuactor which allows sample or probe (stand alone configuration) motion; (3) an electronic system fir detecting and amplificating the signal resulting from the probe-sample interaction; (4) a feedback system allowing to keep the probe-sample interaction stable by the definition of an appropriate setpoint; (5) a system to insulate the microscope from external noise, either mechanical or electrical. The basic operation mode is common to all the scanning probe techniques: a very sharp tip is attached to a cantilever spring which moves over a sample surface. When the tip is moved near the surface, different kind of interactions can take place, ranging from Van Der Waals and dipole–dipole up to magnetic forces. The resulting cantilever spring displacement, acting as a small dynamometer, is recorded as relevant physical signal by an optical lever. A laser beam is focused on the cantilever with an incidence angle θ and reflected via one or more mirrors to a four segment Position Sensitive Detector (PSD), after an optical path L (Fig. 2.10).

Any small variation $\Delta\theta$ of the incidence angle will produce a Δz shift of the reflected spot in the photodiode, amplifying the angle variation as $\Delta z = L\Delta\theta$. This movement is detected by the PSD comparing the intensity of the signal from each segment, with a sub-Angström resolution. Two pairs of segments (upper and lower) give the values of vertical deflection of cantilever, known as topography or Z-signal. Another pair of segments (left and right ones) supplies the so called lateral force signal, related to the tilting of the cantilever due to torsional forces during scan.

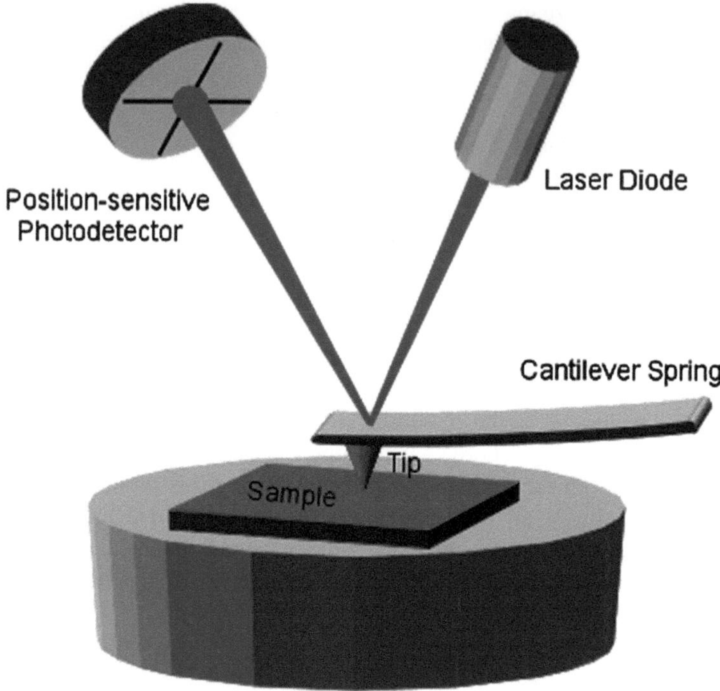

Fig. 2.10 A laser beam is focused on the cantilever and reflected to a PSD

The piezoelectric can contract or elongate under the application of an electric bias with the direction of the elongation that is perpendicular to that of the applied voltage, thus allowing the movement in the x–y–z space. The Z signal goes to the Feedback electronics, which controls the piezoelectric scanner and regulates the interaction strength moving the piezoelectric in order to minimize the difference between the measured value and the required one.

2.2.1.2 The Cantilever

As already mentioned, the probe of an AFM (Fig. 2.11) is the crucial part of the instrument and is made of a sharp tip (with a curvature radius in the order of nanometers) attached to a cantilever type-spring (in the order of microns).

The side of the cantilever opposite to the tip is fixed to a millimiter sized chip, which can be easily handled. The cantilever behaves as a dynamometer, hence subjected to Hook's law, allowing to translate small forces into detectable displacements:

$$F = -k\Delta z$$

Fig. 2.11 Images of different cantilevers taken by scanning electron microscope

The resonance frequency of the cantilever is:

$$w_o = \sqrt{\frac{k}{m}}$$

For a rectangular cantilever the spring constant k is calculated from the equation:

$$k = \frac{Ewt^3}{4l^3}$$

where E is the Young's modulus of the cantilever and l, w and t its length, width and thickness respectively. For example, to obtain a resonance frequency in the range 10–100 kHz with a cantilever spring constant of $\sim$0.1–1 N/m it is necessary a lever approximately 100 μm long and 1 μm thick. Such sizes are nowadays achievable by means of common microfabrication techniques.

To be effective the force sensor has to fulfill several requirements:

(i)   The spring constant k has to be small enough to detect small forces.
(ii)   The resonant frequency w_0 has to be high enough to minimize sensitivity to mechanical vibrations.
(iii) The tip has to be sharp enough to achieve high lateral resolution (of the order of 10^{-10} m).

The tip curvature is the real radius of the tip in the approximation that it is spherical and determines the size of the details that the AFM will be able to resolve. The aspect ratio instead is a measure of the opening angle of the cone representing the full tip, and establishes how deep the tip will penetrate between two adjacent structures on the surface.

2.2.1.3 Operating Modes

Many forces are involved in the interaction between tip and surface. However the most relevant are dipole–dipole attractive interactions and ion core repulsive interaction (viz. Van Der Waals). These forces are described by the 6–12 Lennard–Jones potential:

$$V(r) = C_1 \frac{1}{r^{12}} - C_2 \frac{1}{r^6}$$

where r is the distance from the surface, C_1 and C_2 are appropriate constants, the term to the sixth power of r represents the attractive Van Der Waals potential while the term r^{12} represent the strong repulsive potential when the tip is too close to the surface.

In Fig. 2.12 the Lennard–Jones potential curve is represented.

When the cantilever is far from the surface, the cantilever is in the attractive force regime and the potential varies slowly: the AFM operation mode is called "non-contact mode". A stiff cantilever is forced to oscillate at a certain distance to the sample, without touching it. The forces between the tip and sample are quite low, of the order of pN (10^{-12} N). The detection scheme is based on measuring changes of the resonant frequency or of the oscillation amplitude of the cantilever.

On the contrary, if the cantilevers works in the repulsive force regime, viz. very near to the surface, the operation mode is called "contact mode". This mode enables a very high resolution but it enhances the probability to damage the cantilever. In constant force mode, the tip is constantly adjusted to maintain a constant deflection, and therefore constant distance from the surface. It is this adjustment that is displayed as topographic data. Because the tip is in hard contact with the surface, the stiffness of the lever needs to be less that the effective spring constant holding atoms together, which is on the order of 1–10 nN/nm. Most contact mode levers have a spring constant of <1 N/m.

There is a third operation mode that is "tapping mode", and that is the one used in this thesis.

In this mode the cantilever is excited externally with constant excitation amplitude at a constant frequency near its resonance frequency. While scanning the surface, a feedback loop controls the cantilever-sample distance in order to maintain the amplitude constant. Typical oscillation amplitudes are in the range of 10–100 nm, thus encountering a wide range of tip-sample interactions within each cycle, including both attractive as well as repulsive forces.

The feedback system detects also the oscillation shift (phase signal), which can provide information about the surface composition and viscoelastic properties.

In this thesis a Smena (NT-MDT, Zelenograd, Russia) scanning probe microscope (Fig. 2.13) has been used. This microscope consists of standalone head, where the piezo scanner is positioned inside the head. The geometry of the piezo consists of three blocks of piezoelectric ceramic, they allow the movement of the tip in three orthogonal directions x, y, z. The sample is placed on a special sample holder, and the head is the only moving part. The microscope is equipped with a camera and often with a vibration insulator.

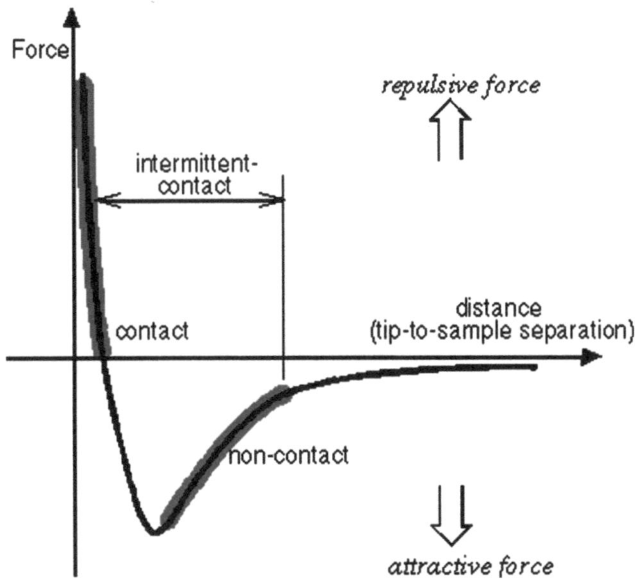

Fig. 2.12 Typical force-curve for a tip-surface interaction

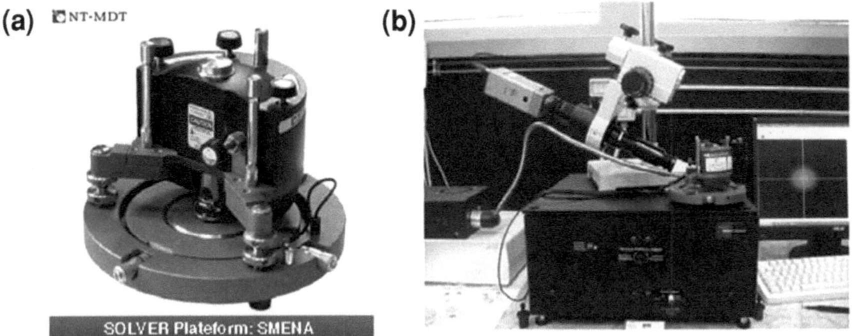

Fig. 2.13 (**a**) NT-MDT^{TM} Smena Standalone head (**b**) AFM equipment (Institute for the study of Nanostructured Materials, CNR-Bologna, Italy)

2.2.2 Scanning Electron Microscopy

2.2.2.1 Working Principle

An electron microscope is a microscope that uses an electron beam for imaging objects, thus allowing an increase in resolution of several orders of magnitude. The development of the electron microscope was based on theoretical work done by Louis de Broglie, who found that wavelength is inversely proportional

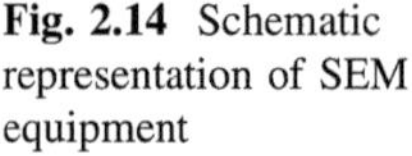

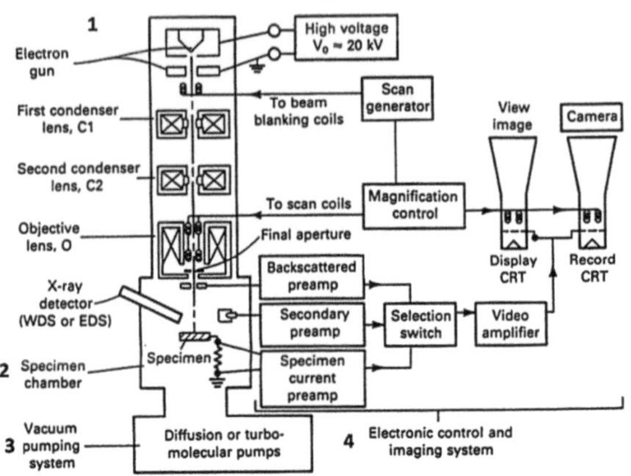

Fig. 2.14 Schematic representation of SEM equipment

to momentum. In 1926, Hans Busch discovered that magnetic fields could act as lenses by causing electron beams to converge to a focus. The first operational electron microscope was presented by Ernst Ruska and Max Knoll in 1932, and 6 years later Ruska had a first version on the market. In 1986 Ruska received a Nobel Prize in physics for his "fundamental work in electron optics and for the design of the first electron microscope". The original form of the electron microscope was a Transmission Electron Microscope (TEM), which uses high energy electrons to image the sample, while Scanning Electron Microscope (SEM) collects secondary electrons to get an image. Figure 2.14 shows the basic components of a SEM: (1) the electron column (2) the specimen chamber (3) the vacuum pumping system and (4) the electron control and imaging system.

In a typical configuration, an electron beam is thermionically emitted from an electron gun fitted with a tungsten filament cathode. Tungsten is normally used in thermionic electron guns because it has the highest melting point and lowest vapour pressure of all metals, thereby allowing it to be heated for electron emission, and because of its low cost. Other types of electron emitters include lanthanum hexaboride (LaB_6) cathodes, which can be used in a standard tungsten filament SEM if the vacuum system is upgraded, and field emission guns (FEG), which may be of the cold-cathode type using tungsten single crystal emitters or the thermally-assisted Schottky type, using emitters of zirconium oxide. The field emission source provides higher resolution; high stability and high current in a small spot size and generates high x-ray fluxes for chemical analysis at high resolution conditions. The electron beam, which typically has an energy ranging from 0.5 to 40 keV, is focused by one or two condenser lenses to a spot about 0.4–5 nm in diameter. The beam passes through pairs of scanning coils or pairs of deflector plates in the electron column, typically in the final lens, which deflect the beam in the x and y axes so that it scans in a raster fashion over a rectangular area of the sample surface.

When the electron beam strikes a sample, these electrons will scatter through the sample within a defined area called the interaction volume (Fig. 2.15).

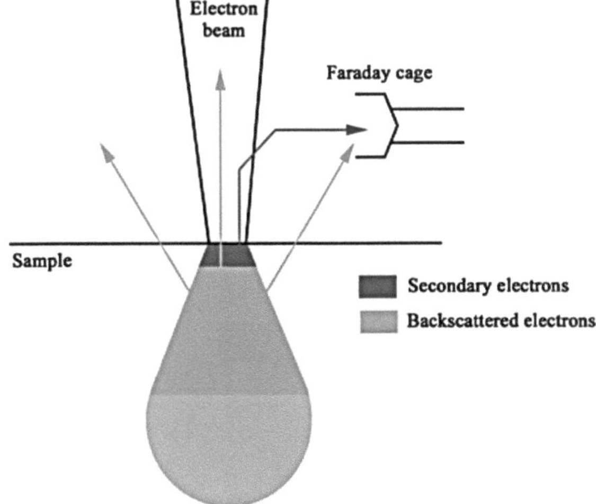

Fig. 2.15 Scheme of the interaction between an accelerated electron beam and the sample

During the electron beam-specimen interactions, many signals are producted, like transmitted electrons (TEM), secondary electrons (SEM), backscattered electrons (BSE) diffracted backscattered electrons (EBSD, that are used to determine crystal structures and orientations of minerals), photons (characteriztic X-rays that are used for elemental analysis and continuum X-rays), visible light (cathodoluminescence-CL), and heat. Secondary electrons are low energy electrons and when produced deeper within the interaction volume, will be absorbed by the sample. Only secondary electrons close to the surface will be able to escape the specimen. The weakly negative secondary electrons will be deflected by a positive pull exerted by the Faraday cage surrounding the secondary electron detector and therefore will contribute to the image formation. Backscattered electrons are also produced deep within the sample but have a much higher energy and because of this, are able to escape from deeper within the interaction volume. Because of their high energy, backscattered electrons will not be deflected by the Faraday cage and therefore not contribute to the image formation. Only a few backscattered electrons will interfere with the signal for secondary electrons. The SEM used all along this work is an Hitachi-S4000 FEG-SEM (Fig. 2.16; max resolution: 1.5 nm; acceleration voltage: 0.5–30 kV; magnification: 20X-200.000X; filament: cold-cathode field emission).

2.2.2.2 Sample Coating

For SEM imaging, samples must be electrically conductive and grounded to prevent electrostatic charge accumulation at the surface. Metal specimens require little preparation apart from a good conventional cleaning procedure. Instead, non-conductive specimens are usually coated with an ultrathin layer (few nm) of an electrically-conducting material (commonly gold, but also platinum, osmium,

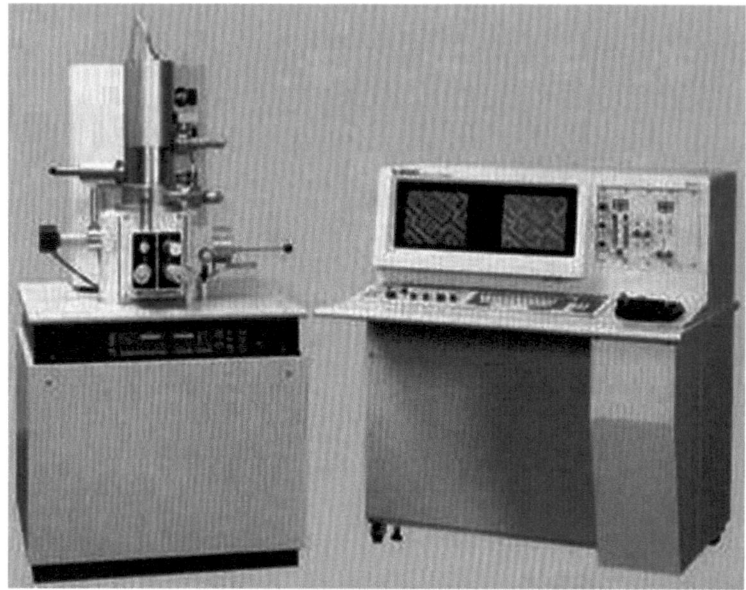

Fig. 2.16 FEG-SEM Hitachi S-4000 (Institute for the study of Nanostructured Materials, CNR-Bologna)

tungsten and graphite) deposited by low vacuum sputtering or high vacuum sublimation. Other than preventing charge accumulation, there are two reasons for coating: to increase signal and surface resolution, especially with samples containing low atomic number elements. The improvement in resolution arises mainly from the enhancing of the backscattering and secondary electron emission near the surface. Depending on the instrument, the resolution can fall somewhere between less than 1 and 20 nm. For SEM analysis, the sample is normally required to be completely dry, since the specimen chamber is at high vacuum; thus living cells and tissues usually require chemical fixation to preserve their structure (see "Materials and Methods" section).

2.2.3 Optical Microscopy

Optical microscopy is a widespread technique that uses UV–visible light to image objects. The resolution limit of optical microscopy is given by the wavelength of the light, so objects spaced less than 0.2 µm are not distinguished by this technique. The simplicity of the technique and the minimal sample preparation required however are significant advantages.

The optical components of a modern microscope (Fig. 2.17) are complex and the whole optical path has to be very accurately set up and controlled. Despite this,

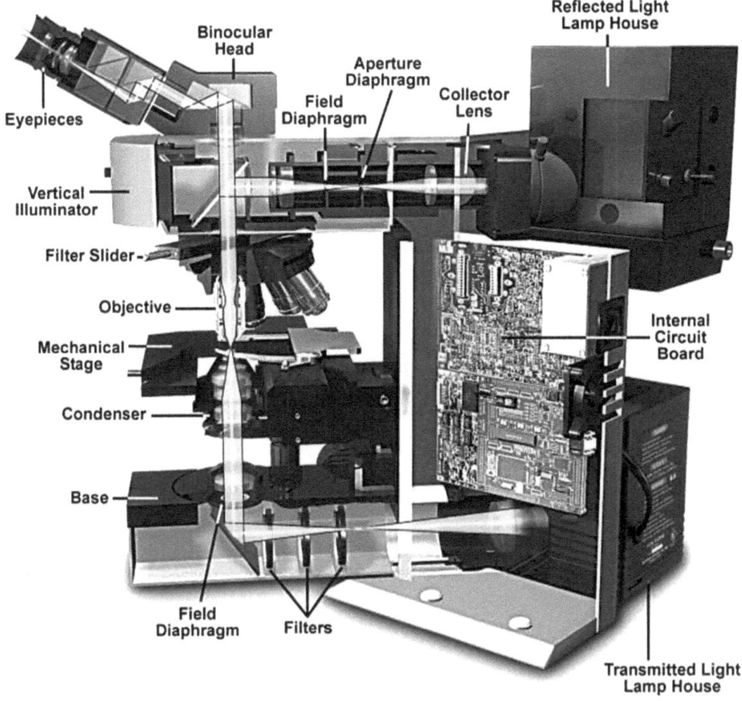

Fig. 2.17 Components of a modern microscope configured for both transmitted and reflected light including two lamp houses, the microscope built-in vertical and base illuminators, condenser, objectives, eyepieces, filters, sliders, collector lenses, field, and aperture diaphragms [30]

the basic operating principle of a microscope is quite simple. An objective lens with very short focal length (few millimeters) is used to form a highly magnified real image of the object. This magnification is carried out by means of two lenses: the objective lens which creates an image at infinity, and a second weak tube lens which forms a real image in its focal plane. The resulting image can be detected directly by the eye, imaged on a photographic plate or captured digitally.

In the optical microscopy field, several techniques working with different illumination modes have been developed such as Bright field, Dark field, Phase contrast, Polarized light, Hoffman modulation contrast, Differential interference contrast or Fluorescence [30].

Bright field microscopy is the simplest, with the sample being illuminated by transmitted white light. It presents however some limitations such as it can only image dark or strongly refracting objects effectively. Darkfield microscopy describes an illumination technique used to enhance the contrast in unstained samples. To view a specimen in dark field, an opaque disc is placed underneath the condenser lens, so that only light that is scattered by objects on the slide can reach the eye. Instead of coming up through the specimen, the light is reflected by

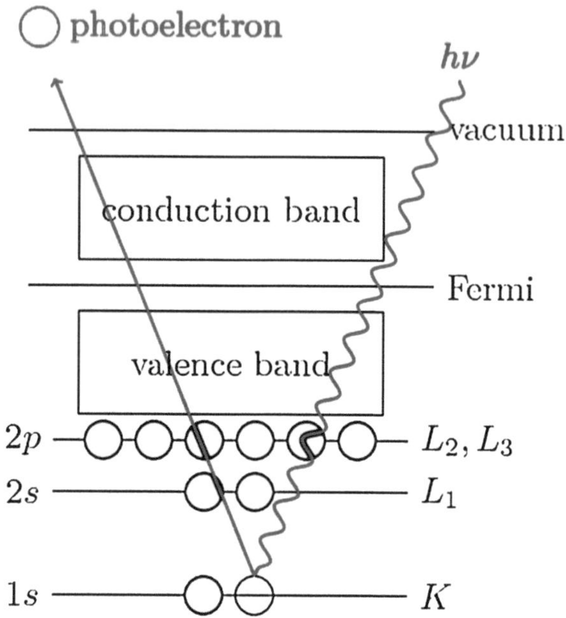

Fig. 2.18 The photoelectron is emitted as a consequence of the irradiation by a high energy photon

particles on the slide. This produces the classic appearance of a dark, almost black, background with bright objects on it.

In this work, optical microphotographs have been recorded both in bright/darkfield and in fluorescence mode with a Nikon 80i (Tokio, Japan) microscope equipped with an epi-illuminator, a dark field system and cross polars, using $50\times$ and $100\times$ objective and a digital camera (Nikon Digital sight DS-2Mv Japan).

2.2.4 X-Ray Photoelectron Spectroscopy

The analysis of solid surfaces is one of the most intensively developing fields in present-day analytical chemistry. This is mainly due to the need for the qualitative and quantitative determination of the elemental composition of surfaces. X-ray Photoelectron Spectroscopy (also known as Electron Spectroscopy for Chemical Analysis, ESCA) is based on the photoelectric effect, which consists in plain terms in the emission of electrons by the filled core states of a solid upon its irradiation with electromagnetic waves (Fig. 2.18).

The photoelectric effect was discovered by the well-known German physicist G. Hertz in 1887 and theoretically explained by Einstein in 1905. Instruments for recording X-ray photoelectron spectra and the technology of high vacuum were rapidly developed in the next decades. By the mid-1950s, the experimental and theoretical grounds were laid for the first spectroscopic method for surface analysis, X-ray photoelectron spectroscopy. The breakthrough was made in 1954, when a group of Swedish scientists headed by the Professor K. Siegbahn (Nobel Prize 1981

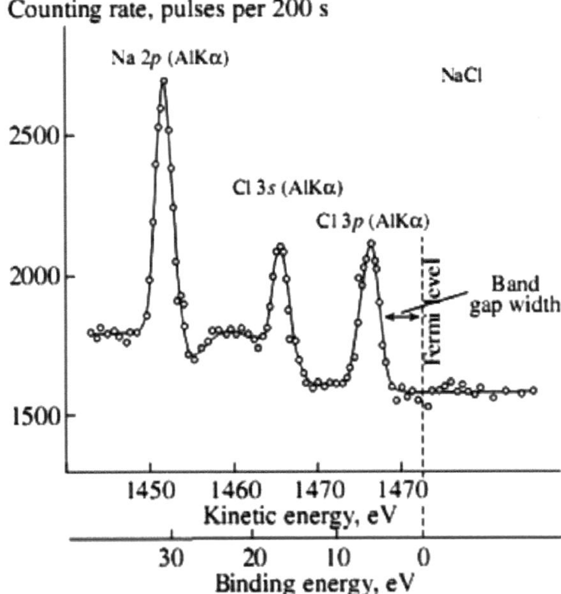

Fig. 2.19 First XPS spectrum of sodium chloride (1954)

in physics) proposed the first high-resolution electron spectrometer for determining slow electrons forming upon the irradiation of a solid surface with X-rays. Using this instrument, they first observed separate characteriztic peaks in the X-ray photo-electron spectrum of sodium chloride surface (Fig. 2.19).

It was found that the proposed method gives analytical information from very small depths (3–5 nm) and can be successfully used for determining the elemental composition of solid surfaces.

Nowadays in XPS the sample is irradiated with a monochromatic X-ray beam (1.2 keV $< hv <$ 1.4 keV) causing photoelectrons to be emitted from the surface; essentially all XPS measurements are done exploiting the Al K_α-line ($hv =$ 1486.6 eV) or Mg K_α-line ($hv =$ 1253,6 eV) as beam source. An energy-resolved electron detector with high energy resolution $E_{kin}/\Delta E$ and large collection angle (focusing ability) determines the binding energy (E_B) of the emitted photoelectrons (Fig. 2.20).

The fundamental equation of the photoelectric effect is the following:

$$E_{kin} = hv - E_B - \Phi$$

Where E_{kin} is the kinetic energy of the photonelectron, hv the energy of the incident photon and Φ is the workfunction value of the material irradiated. In Fig. 2.21 I report a schematic representation of an XPS spectrum (primary structure), with the binding energy in x-axis and the intensity of the signal in y-axis.

From the binding energy and intensity of a photoelectron peak, the elemental identity, chemical state, and quantity of an element can be determined.

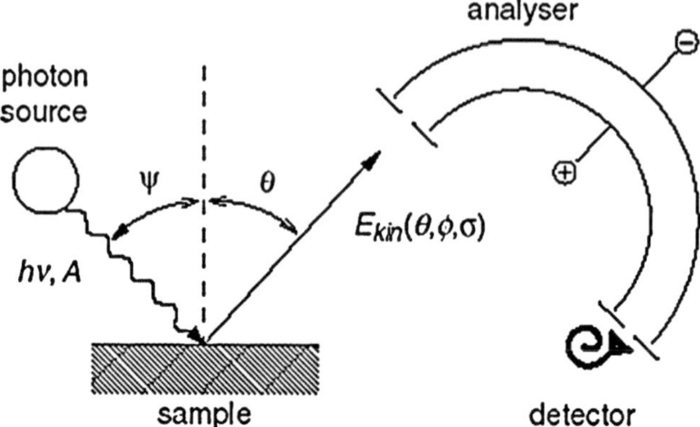

Fig. 2.20 XPS basic principle

Fig. 2.21 Elementary representation of a primary XPS spectrum

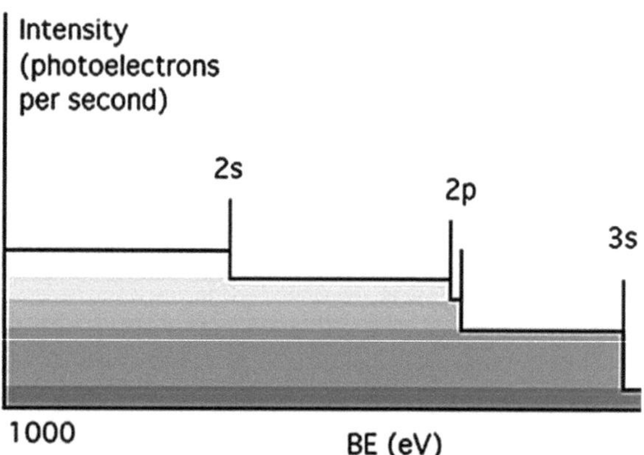

The information that XPS provides about surface layers or thin film structures is of value in many industrial applications including: polymer surface modification, catalysis, corrosion, adhesion, semiconductor and dielectric materials, electronics packaging, magnetic media, and thin film coatings used in a number of industries.

The method is being steadily improved on the basis of recent achievements in science and technology. In particular, the use of synchrotron radiation for the excitation of photoelectron emission could significantly improve the spectral resolution of the method and lower its detection limit. One of the latest methods is X-ray photoelectron microscopy, in which the spatial distribution of elements over a surface can be studied using characteriztic photoelectrons.

In my work, XPS measurements were carried out with an ESCALAB MkII (VG Scientific Ltd., U.K.) spectrometer equipped with a standard Al KR excitation source and a 5-channel detection system (Institute for the study of Nanostructured Materials, CNR-Rome).

References

1. Xia Y, Rogers JA, Paul KE, Whitesides GM (1999) Chem Rev 99:1823–1848
2. Sgarbi N, Pisignano D, Di Benedetto F, Gigli G, Cingolani R, Rinaldi R (2004) Biomaterials 25:1349–1353
3. Bystrenova E, Facchini M, Cavallini M, Cacace MG, Biscarini F (2006) Angew Chem Int Ed Engl 45:4779–4782
4. Martin CR, Aksay IA (2004) J Electroceram 12:53–68
5. Beh WS, Kim IT, Qin D, Xia YN, Whitesides GM (1999) Adv Mater 11:1038–1041
6. Kumar A, Whitesides GM (1993) Abstracts of Papers of the American Chemical Society, 206, 172-COLL
7. Xia YN, Kim E, Zhao XM, Rogers JA, Prentiss M, Whitesides GM (1996) Science 273:347–349
8. Zhao XM, Xia YN, Whitesides GM (1997) J Mater Chem 7:1069–1074
9. Kim E, Xia YN, Whitesides GM (1996) J Am Chem Soc 118:5722–5731
10. Xia YN, Whitesides GM (1998) Annu Rev Mater Sci 28:153–184
11. Kumar A, Whitesides GM (1993) Appl Phys Lett 63:2002–2004
12. McDonald JC, Whitesides GM (2002) Acc Chem Res 35:491–499
13. Zhao XM, Stoddart A, Smith SP, Kim E, Xia Y, Prentiss M, Whitesides GM (1996) Adv Mater 8:420
14. Schueller OJA, Brittain ST, Marzolin C, Whitesides GM (1997) Chem Mat 9:1399–1406
15. Trau M, Yao N, Kim E, Xia Y, Whitesides GM, Aksay IA (1997) Nature 390:674–676
16. Delamarche E, Bernard A, Schmid H, Michel B, Biebuyck H (1997) Science 276:779–781
17. Kim E, Whitesides GM (1997) J Phys Chem B 101:855–863
18. Huang WF, Liu QS, Li Y (2006) Chem Eng Technol 29:716–723
19. Cavallini M, Biscarini F (2003) Nano Lett 3:1269–1271
20. Wyart FB, Daillant J (1990) Can J Phys 68:1084–1088
21. Sharma A, Khanna R (1999) J Chem Phys 110:4929–4936
22. Xie R, Karim A, Douglas JF, Han CC, Weiss RA (1998) Phys Rev Lett 81:1251–1254
23. Cavallini M, Facchini M, Albonetti C, Biscarini F (2008) Phys Chem Chem Phys 10:784–793
24. Suematsu NJ, Ogawa Y, Yamamoto Y, Yamaguchi T (2007) J Colloid Interface Sci 310:648–652
25. Cavallini M, Albonetti C, Biscarini F (2009) Adv Mater 21:1043–1053
26. Collier CP, Wong EW, Belohradsky M, Raymo FM, Stoddart JF, Kuekes PJ, Williams RS, Heath JR (1999) Science 285:391–394
27. Cavallini M, Calo A, Stoliar P, Kengne JC, Martins S, Matacotta FC, Quist F, Gbabode G, Dumont N, Geerts YH, Biscarini F (2009) Adv Mater 21:4688
28. Cavallini M, Bergenti I, Milita S, Ruani G, Salitros I, Qu ZR, Chandrasekar R, Ruben M (2008) Angew Chem Int Edit 47:8596–8600
29. Cavallini M, Facchini M, Massi M, Biscarini F (2004) Synth Met 146:283–286
30. Betzig Eric, Patterson George H, Sougrat Rachid et al (2006) Imaging intracellular fluorescent proteins at nanometer resolution. Science 313:1642–1645

Chapter 3
Stable Non-Covalent Functionalization of Teflon-AF

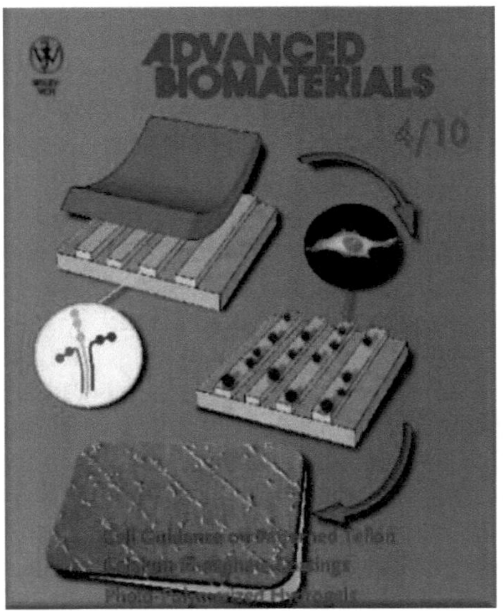

3.1 Introduction

A straightforward approach for the control of cell fate is based on the fabrication of neighboring surface features exhibiting opposite cell anchoring properties. Cell growth and migration can be guided through the fabrication of a pattern of antifouling regions coexisting with regions favouring cell growth [1]. Several materials with antifouling properties have been employed to guide cell adhesion

M. Bianchi, *Multiscale Fabrication of Functional Materials for Regenerative Medicine*, Springer Theses, DOI: 10.1007/978-3-642-22881-0_3,
© Springer-Verlag Berlin Heidelberg 2011

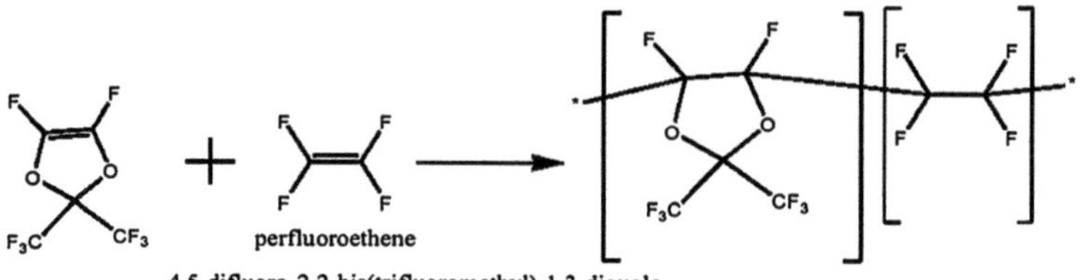

Fig. 3.1 Addition polymerization reaction for Teflon-AF synthesis

and migration such as PEG [2], lipids [3], thermo-responsive polymers based on PNIPAAm [4].

A technologically interesting material, yet not extensively investigated as non-fouling surface, is polytetrafluoroethylene (PTFE, Teflon) and its copolymers, such as Teflon-AF (Poly[4,5-difluoro-2.2-bis(trifluoromethyl)-1,3-dioxole-cotetrafluoroethylene], Fig. 3.1).

Teflon is a biocompatible, inert and cell-non-adhesive material, employed as safety material for medical tools and instruments [5]. Teflon is also relevant in biosensing for its dielectric properties, as for instance in sensors based on dual gate organic field effect transistors (OFET) [6–8]. Cells have a very low adhesion propensity on Teflon, and the rare adhesion events are often observed in correspondence of large defects. Few reports in literature have addressed the use of Teflon as substrate for patterning and controlling cell adhesion [9–11]. Teflon-derivates as Teflon-AF could become an interesting material for cell guidance, cell-carrier, or scaffold if selectively patterned with cell-adhesion biomolecules. A stable functionalization of Teflon-AF surface with molecules promoting cell adhesion by non covalent attachment and no chemical modification of the Teflon-AF surface is challenging.

Standard thin film deposition techniques, as drop casting and spin coating, are not effective because the extreme hydrophobicity of Teflon-AF prevents a proper wetting of the surface by aqueous solutions. Moreover, surface roughness and defects may pin the solution leading to inhomogeneous deposits from droplets. In Chap. 2, I showed that LCW technique is based on the pinning of the solution between the surface and the protrusions of the elastomeric stamp, placed on top of the solution, which leads to the formation of menisci. The deposition of the solute on the underlying surface is confined within the menisci, whose shape and the lateral size are related to the surface tension of the surface and of the stamps [12]. When the substrate surface is highly hydrophobic, such as that of Teflon-AF the menisci either do not form, or become unstable, preventing the aqueous solution from appropriately wetting the substrate. It is therefore necessary to change the surface tension of the stamp in order to obtain stable menisci.

As material with fouling properties, I chose laminin (Fig. 3.2).

Laminin is the principal non-collagenic component and biologically active part of the basal lamina, a protein network foundation for most cells and organs [13]. More specifically laminin is a trimeric protein that contains an α-chain, a β-chain,

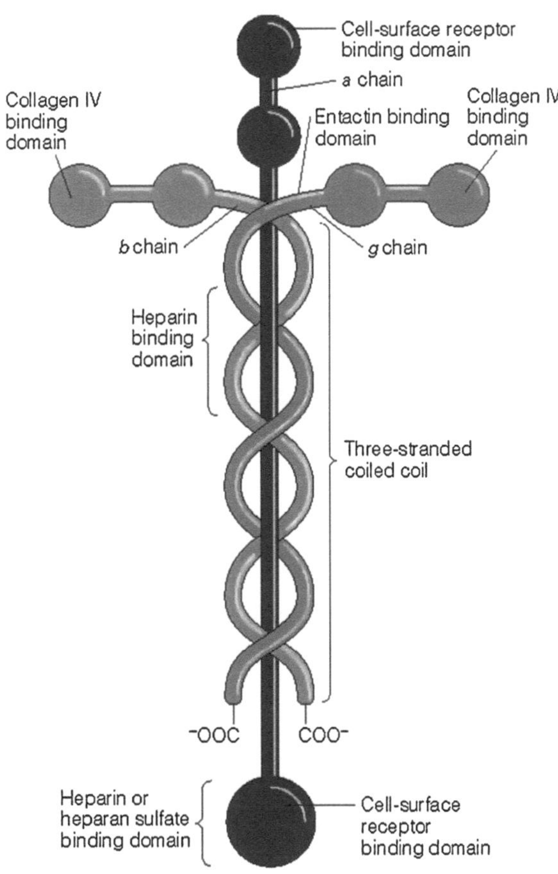

Fig. 3.2 Schematic representation of laminin protein, evidencing the trimeric structure

and a γ-chain, found in five, three, and three genetic variants, respectively. The trimeric proteins form a plus sign with one long arm, giving a structure that can bind to other cell membrane and extracellular matrix molecules. The three shorter arms are able to bind to other laminin molecules, which allow them to form sheets. The long arm is capable of binding to cells, which helps anchor organized tissue cells to the membrane.

I demonstrate here that it is possible to achieve a local functionalization of Teflon-AF surface with laminin, by exploiting the modulation of the stamp hydrophobicity, thus providing a stable anchoring and growing of neural cells. Cell adhesion on these patterned substrates is highly selective, and cell confluence can be achieved exclusively on the laminin patterned regions.

3.2 Laminin Functionalization of Teflon-AF

Teflon-AF films have been obtained by a sol-gel method as described in Sect. 3.5. The functionalization of these Teflon-AF films with laminin is realized by means of LCW-plasma modified technique. The polydimethylsiloxane (PDMS) mold is placed on top of a droplet (5 μl) of laminin solution previously cast onto the Teflon-AF coated glass. The solution is left to evaporate and the stamp is then gently peeled-off. The final pattern of laminin faithfully reproduces the channels of the PDMS stamp. Areas as large as 50 mm^2 are patterned with a homogeneous motif displaying features over two different length scales: channels 17–70 μm wide and 1.5 μm high, channels 900 nm wide and 150 nm high; the former have been obtained by replica molding of a master obtained by UV-photolithography (see Sect. 3.5), the latter starting from a commercial Compact Disk master. These figures of merit would not be possibly achieved using two extremely hydrophobic surfaces as PDMS and Teflon-AF. Indeed a combination of hydrophobic surfaces like these would generate unstable menisci with high contact angle values, resulting into an irregular and poorly resolved pattern (Fig. 3.3d) with the laminin dispersed over all the surface and some uncontrolled excess of concentration of the deposit corresponding to the channels of the stamp.

Upon these conditions, despite of the presence of laminin which may help the cell attachment to the surface, spatially directed cell adhesion would be hindered by the absence of a well defined alternance of pro- and anti-adhesion regions (Fig. 3.3g). In order to obtain a regular pattern the meniscus must be pinned steadily at the stamp surface by increasing its surface tension.

For this reason I have carried out an oxygen plasma treatment of the bottom of the stamp. The bare PDMS surface has a contact angle with pure water of approximately 105° (Fig. 3.3a) while upon the O$_2$ plasma treatment extent, it displays values ranging between 15° and 40° (Fig. 3.3b and c).

Compared to the bare PDMS stamps, the LCW performed with more hydrophilic channels leads to well defined and well separated stripes over all the length scales relevant for the cell adhesion process: from hundred nanometers (comparable to the length scale of adhesion points within a single cell), up to tens of micrometers (Fig. 3.3e) to accommodate one or more cells and direct their growth (Fig. 3.3h). Under conditions where an extremely hydrophilic stamp is obtained, the amount of protein released on the surface decreases leading to a pattern where the stripes becomes segments (Fig. 3.3f) with a consequent inhomogeneous cell distribution (Fig. 3.3i).

In Fig. 3.4 a schematic representation of the plasma modified LCW explains how the pattern transfer takes place: the increased wettability of the more hydrophilic PDMS surface upon the plasma treatment leads to a confinement of the solution droplets within the channels, and the formation of stable menisci with a lower contact angle at the PDMS surface. This configuration is different from the LCW originally developed by Cavallini and Biscarini [12] where pinning occurred at protrusions, rather than channels. Here the lack of available contact surface at

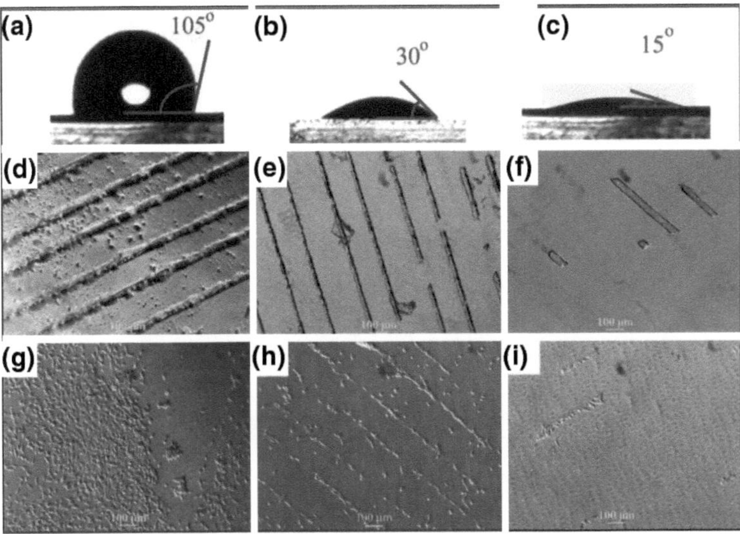

Fig. 3.3 Optical images of water droplets with their static contact angles on bare PDMS (**a**) and on the same material exposed to oxygen plasma 2 min (**b**) and 5 min (**c**); (**d**), (**e**) and (**f**) are optical images (bright field) of the laminin patterns obtained with the corresponding PDMS stamps as in (**a**), (**b**) and (**c**) on Teflon-AF surfaces are shown. (**g**), (**h**) and (**i**) report the cell distribution on (**d**), (**e**) and (**f**) substrate, respectively, upon 24 h incubation

the protrusions is compensated by the increase of surface area of the recesses, provided the solvent (water in this case) infills them. Thus, the aspect ratio of the recesses (together with the duty cycle of the pattern which is the ratio between the projected area of the protrusions and the one of the recesses) becomes an important parameter of LCW control together with the surface tension.

Figure 3.5a shows an AFM image of a Teflon-AF film obtained by spin coating with a representative section analysis reported in Fig. 3.5c. The surfaces obtained by this protocol are very flat as confirmed by the measured roughness (rms roughness = 1.70 nm). Figure 3.5b and d shows the AFM image and the respective height profile of a regular laminin pattern, 1,000 nm wide stripes separated by 500 nm bare Teflon-AF, realized by LCW using a PDMS replica of a recordable Compact Disc. To the best of our knowledge this is the first reported evidence of a direct patterning of an extremely hydrophobic surface such as the Teflon-AF. The edge along the stripes is sharp, revealing well-defined lines of laminin whose height (150 nm) can be tuned by the protein concentration in the depositing solution. The shape of the grooves in the profile suggests that the envelope of the tip might be present although simple considerations [14] based on the tip geometry used lead us to conclude that the measured depth of the recesses is not limited by the tip size.

Interestingly, the pattern feature height can be tailored by changing the protein concentration in the starting solution. In Fig. 3.6, AFM images of Teflon-AF

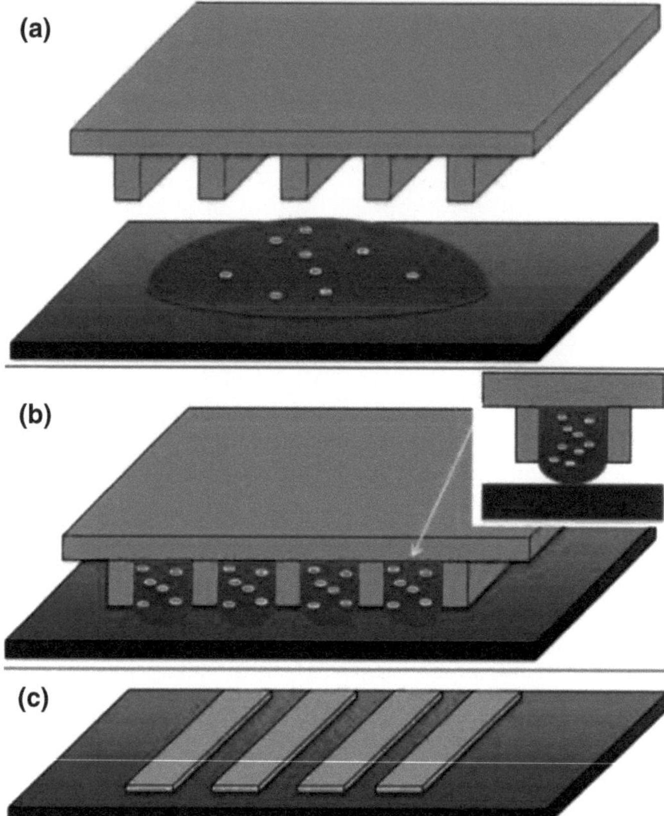

Fig. 3.4 LCW process realized using an oxygen plasma treated PDMS stamp exhibiting a pattern of protrusions and recesses. (**a**) A drop of laminin solution is placed on the Teflon-AF surface; (**b**) the elastomeric stamp is then placed on top of the solution which infills the recesses (hydrophilic channels); (**c**) upon solvent evaporation the solute is deposited reproducing the geometric pattern of the recesses. In (**d**) the solution droplet formed between the hydrophilic plasma treated PDMS channel and the hydrophobic Teflon-AF surface is reported

surfaces patterned by LCW with two different laminin concentrations using a CD replica as stamp, are reported.

In Fig. 3.6, an AFM image of the pattern (a) is reported along with a line profile (c) for a 20 μg ml^{-1} laminin in DMEM. The height of the features is 240 $\pm$ 10 nm. In (b) and (d) the same data are reported for a 2 μg ml^{-1} laminin sample. In this case the height of the features is 137 $\pm$ 10 nm.

Hence it is possible to achieve different pattern feature heights simply by starting from different laminin concentrations.

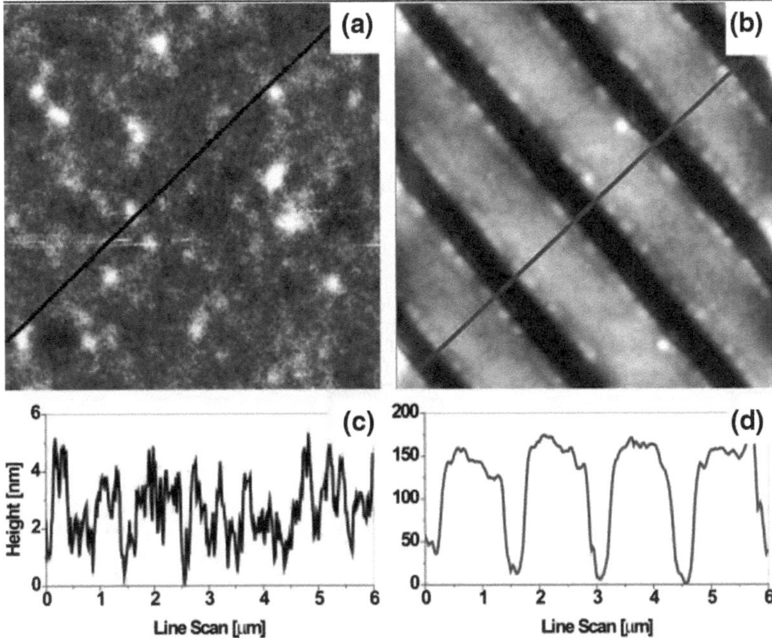

Fig. 3.5 (**a**) 5 × 5 μm² AFM image of the bare Teflon-AF surface with a representative height profile (**c**); (**b**) the same surface after laminin patterning by LCW performed using a Compact Disk PDMS replica as stamp. The line profile in (**d**) shows the characteriztic sizes of the patterned laminin stripes

3.3 Laminin Immunofluorescence Assay

The local distribution of laminin on the substrate and the specific functional binding of the patterned protein have been evaluated by immunofluorescence assay. The choice of pattern design allows one to functionalize regions of different size and shape. In Fig. 3.7a is shown an optical image of a laminin pattern realized starting from a PDMS stamp with channels of variable width in the range between 17 and 70 μm.

The larger stripes can accommodate up to 5–6 cells across, whereas the narrower ones only a single cell. The stripes are spaced apart by a distance considerably larger than the cell diameter in order to demonstrate that cell adhesion is dictated by preferential interactions with the laminin template and not by simple geometrical constraints.

The non covalent interactions between the protein pattern (Fig. 3.7a) and the inert Teflon-AF surface have made necessary to assess the effect of the dissolution of the deposited laminin in the culture media under the cell culture conditions (minimum 24 h in incubator). Hence, immunofluorescence staining has been performed upon incubation of the substrates for 24 h in complete cell culture

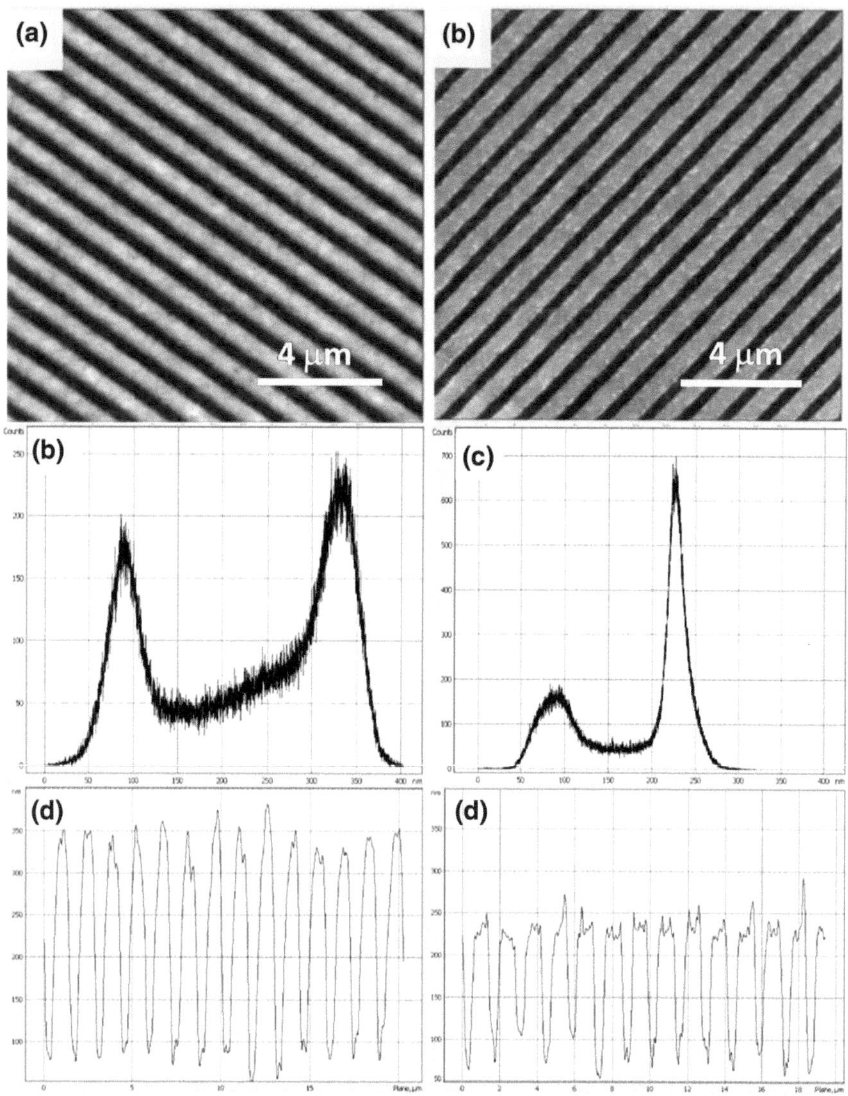

Fig. 3.6 20 μm AFM images and respective line profiles showing stripe height modulation obtained by using two different protein concentrations: 20 μg ml^{-1} (**a, c**) and 2 μg ml^{-1} (**b, d**)

medium at 37°C and controlled humidity. Figure 3.7b clearly shows how the proteins are located along the channels and display their specific binding during incubation time intervals compatible with the cell adhesion and proliferation events. It is worthwhile underlining that the spatially controlled interaction between the laminin and its specific antibody demonstrates how this substrate could be employed for biosensors to detect protein–protein interactions.

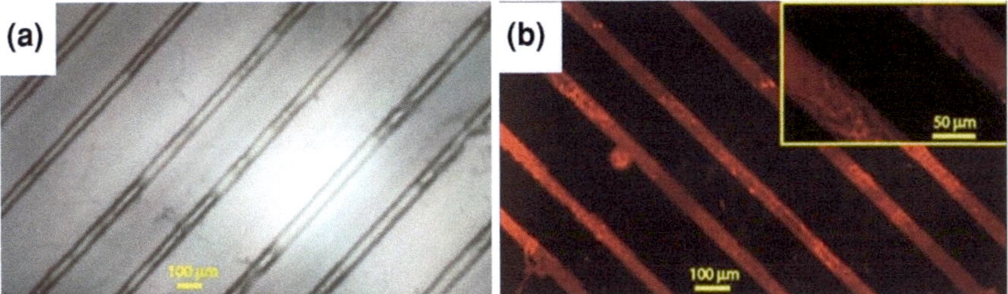

Fig. 3.7 (**a**) Optical image (bright field) of a pattern of laminin with channel width ranging between 17 and 70 μm and its relative immunofluorescence staining with anti-laminin (**b**) upon incubation for 24 h under the conditions used for cell culture (DMEM-F12, 37°C); the higher magnification image reported in the inset allows appreciating the strong localization of laminin

Remarkably, such a stable controlled functionalization of the substrate has been achieved without any chemical modification of the Teflon-AF substrate in a one-step procedure.

3.4 Neural Cell Growth on Laminin-Patterned Teflon-AF

To monitor the effectiveness of the functionalization on the control of the cell adhesion and growth human SH-SY5Y neuroblastoma cells have been seeded on the patterned surfaces and allowed to proliferate for several days. I have fabricated a pattern suitable to investigate the selective adhesion on the functionalizated surface areas displaying interrupted stripes as those reported in Fig. 3.8a.

I used a non-continuous line in order to demonstrate that confinement of cell-growth can be achieved both laterally and longitudinally. This particular geometry has been obtained using the same continuos line master but changing the surface tension of the PDMS replica; the amount of laminin pattern released onto the Teflon-AF surface has been controlled by the hydrophilicity of the stamp as reported in Fig. 3.3. In Fig. 3.8b and 3.8c it is possible to observe the highly selective neuroblastoma cell adhesion after 24 and 48 h of incubation. The pattern is reproduced with complete conformality. Independently on the segment length and lateral size, upon 48 h of incubation, cells have adopted a more "spread-out" morphology with migration of some cells into non-printed areas. This proves that laminin pattern is an excellent template for inducing highly preferential, if not selective, cell adhesion. The cells remain prevalently confined on laminin areas and viable even during longer period (5 days, Fig. 3.9).

This behaviour is due to the strong gradient of adhesive properties between the antifouling Teflon-AF areas and the adhesion promoting laminin. This can be stressed by the results shown in Fig. 3.10 that reports the boundary region between a large laminin square and the neighboring polyfluorinated surface.

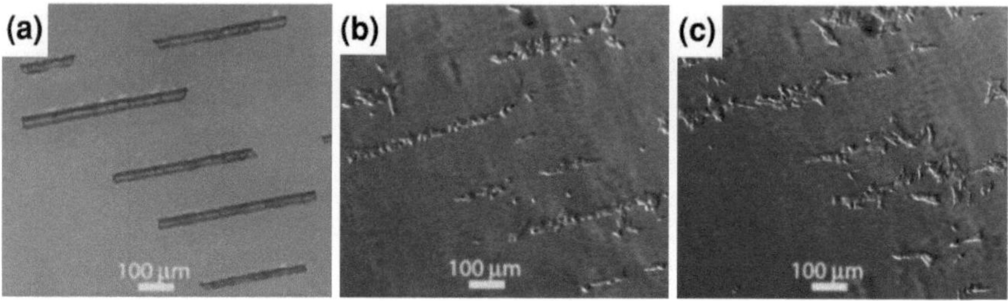

Fig. 3.8 (**a**) Pattern of laminin and the same region after seeding SH-SY5Y neuroblastoma cells. At 24 h the cells are confined on the laminin pattern (**b**) and they reach confluence only on the patterned regions (**c**) after 48 h

Fig. 3.9 SH-SY5Y neuroblastoma cells upon 5 days incubation on the Laminin patterned Teflon surface

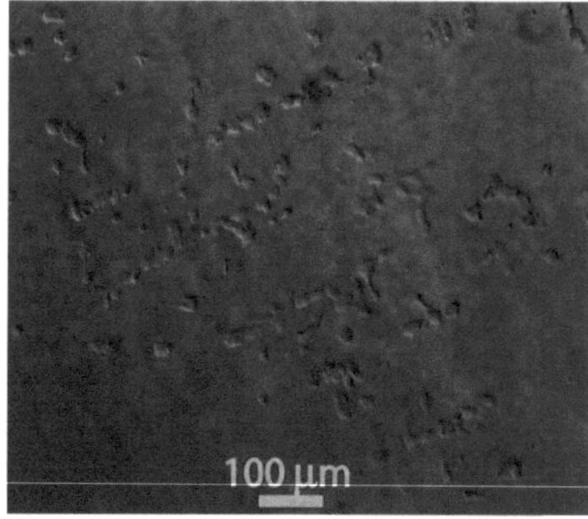

Fig. 3.10 Optical image (bright field) of the border between a bare Teflon-AF region (*top* of the area) and a laminin functionalized region (*bottom*). The cells are strictly confined within the laminin functionalized region

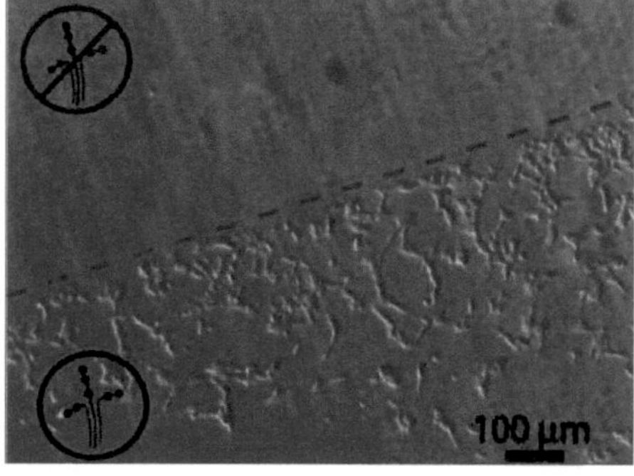

Cells grow only on the laminin pattern while they do not adhere on the outer Teflon-AF surface at all. The preferential adhesion, and hence the guidance by the template, remains evident also for prolonged time of incubation (48 and 72 h). These results suggest that the dissolution of laminin from the printed areas occurs via desorption from the surface into the media rather than via lateral diffusivity and re-anchoring.

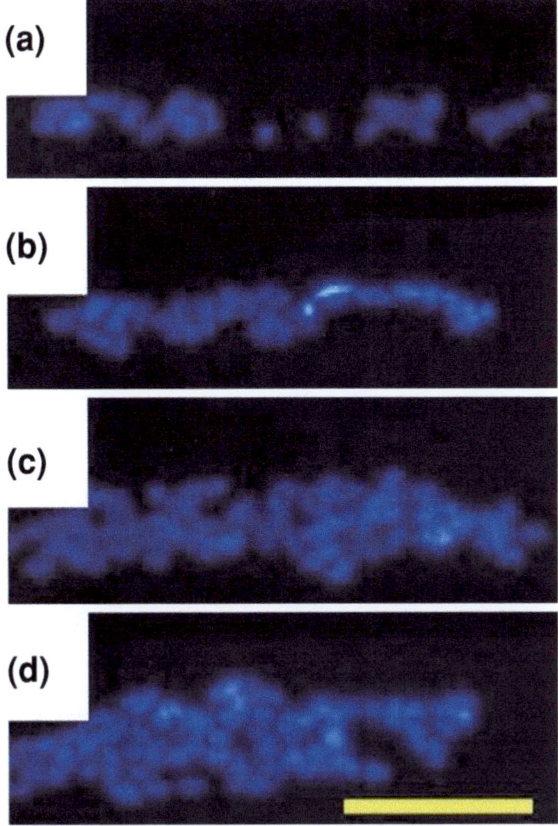

Fig. 3.11 Neuroblastoma nuclei stained with DAPI (4′,6-diamidino-2-phenylindole) attached onto laminin stripes of increasing width (**a–d**)

As mentioned before, patterns were designed in order to accommodate from 1 to 5–6 cells across. In Fig. 3.11 the distribution of DAPI-stained SH-SY5Y cell nuclei on different-width laminin stripes at 48 h from seeding is shown.

From the immunofluorescence images, it is possible to notice that the larger stripe may host up to 6 cells across while the thinner ones form colonies close to a single cell lane, enabling a fine control on the cell density distribution.

3.5 Materials and Methods

3.5.1 Preparation of Telfon-AF Films

Microscope glass supports (Thermoglass S.r.l., 1.5 cm × 1.5 cm) were cleaned by rinsing in piranha solution (H_2SO_4:H_2O_2, 3:1 v/v) at 100°C for 15 min. Teflon-AF (Sigma Aldrich) was dissolved in Fluorinert FC-40 (Sigma Aldrich) at concentration of 16.5 mg ml^{-1} at room temperature; the solution was then spin-coated onto the glass substrates by a Laurell Technologies Corporation Spin Coater (30″, acc. 100, 1500 rpm.). The final thickness of the film was around 150 nm, as supported by AFM analysis.

3.5.2 Pattern Fabrication on PTFE Surface

A PDMS mold was obtained by replica molding of a master with designed topographical features (microchannels). The width of the channels range from one to several cell diameters. The distance between adjacent channels was kept unvaried, thus the coverage of the final pattern is varying linearly across the sample. The aspect ratio of the microchannels was kept intentionally low (approximately 0.1). Although low aspect ratio may induce sagging effects of the PDMS replica, protein stripes obtained after functionalization are smooth and the onset of Rayleigh instabilities during patterning is reduced. The master was fabricated by contact photolithography with Mask Aligner Karl Suss MJB40 ($\lambda = 365$ nm). High contrast masks were obtained from photoreduction of initial inkjet printout. The exposure interval was adjusted to match the contrast of the mask. Negative photoresist AR P3210 (All Resist), was spin coated at 2000 RPM on polycarbonate flat substrate (surface roughness <10 nm), resulting in a film with thickness 5 μm. After alignment and exposure, the master was developed for 1 min into developer AR 300-35 and postbaked at 150°C for 30 min, to enhance crosslinking of photoresist. During postbaking, the section of the channel becomes round shaped; this ensures the peel-off of the mold in replica molding. The patterned areas were always in the range of 5 × 5 mm. PDMS (Sylgard 184) and its curing agent were mixed in ratio 10:1, degassed and poured on top of the polycarbonate/AR P3210 master. Curing was performed at 80°C for 8 h, in standard thermostatic oven. After peel-off, the PDMS replica were rinsed and stored in deionized water. In order to increase their hydrophilicity the PDMS stamps were treated with oxygen plasma at constant pressure ($O_2 = 40$ mTorr) and power (40 mW). By varying the time of exposure to O_2 plasma we tune the surface tension of the stamp. The PDMS mold was lately placed on top of a droplet (5 μl) of Laminin solution (20 μg ml^{-1} in cell culture medium) previously pipetted onto the PTFE coated glass. The solution was left to evaporate for 1 h under ambient conditions and the stamp was then gently peeled-off. The final pattern of Laminin faithfully reproduces the channels within the PDMS stamp.

3.5.3 Characterization of the Patterned Substrates

The patterns of Laminin were characterized from the morphological point of view by optical and SFM imaging. Optical images were collected by an Olympus IX70 in transmission mode with 10X objective. SFM images were acquired with a Smena microscope (NT-MDT, Moscow, Russia) operated in semi-contact mode in ambient conditions. The cantilever employed were NSG (NT-MDT, Moscow, Russia) with a nominal tip radius of curvature 10 nm and a resonance frequency between 90 and 230 kHz. Images were analyzed by using the free software Image Analysis (NT-MDT, Moscow, Russia); the size and the height of the pattern features were measured by the section analysis tool. The local distribution of laminin on the substrate and the specific functional binding of the patterned protein were evaluated by immunofluorescence assay. Laminin patterned.

Teflon-AF samples were blocked for 30 min with phosphate buffer saline (PBS) 1X containing 1% bovine serum albumin (blocking solution). They were incubated for 1 h at room temperature with antilaminin antibody (L6274 Sigma Catalog 2010) diluted 1:25 in blocking solution, washed three times with PBS 1X, and incubated for 90 min at room temperature with the fluorescent Alexa Fluor 594 antirabbit antibody (Invitrogen) diluted 1:200 in PBS 1X. Control samples were incubated in blocking solution with no primary antibody. After washing with PBS 1X, all samples were analyzed under an epifluorescence microscopy (Olympus IX70). Some experiments were carried out on laminin patterned Teflon-AF substrates, previously maintained in the presence of cell culture medium in standard culture conditions (37°C in a humidified atmosphere with 5% CO_2) for 24 h.

3.5.4 Testing of the Cell Adhesion and Growth

Human SH-SY5Y neuroblastoma cells (purchased from European Collection of Cell Cultures, ECACC, Salisbury, UK) were cultured in an 1:1 mixture of F-12 nutrient mixtures (Ham12) and EMEM, supplemented with 15% foetal bovine serum (FBS), 2 mM L-glutamine, 1% non-essential amino acids, penicillin (100 U ml^{-1}), and streptomycin (100 mg ml^{-1}) (complete medium). Cells were maintained under standard culture conditions and fed every 2–3 days.

In order to evaluate the cell adhesion and growth on laminin patterned substrates, neuroblastoma cells were seeded at the density of 3000 cells cm^{-2} on bare and laminin patterned Teflon-AF substrates in a final volume of 400 ml into a PDMS pool. The samples were maintained in standard culture conditions for different times. Then, the cell nuclei were visualized by using an epifluorescence microscopy after incubation of the samples with DAPI (4′,6-diamidino-2-phenylindole) (dil. 1.500 in PBS 1X) for 5 min at room temperature.

3.6 Conclusions

I have performed a spatially controlled surface functionalization of strongly hydrophobic and anti-adhesive Teflon-AF substrates by means of a modified LCW patterning which exploits the confinement of the solution of interest between the Teflon-AF surface and the stamp recesses. Such a confinement has been controlled by the differential wettability of the stamp and the substrate achieved throughout a surface tension modification of the PDMS stamps by O_2 plasma treatment. A non-covalent modification of the surface has been possible without any chemical or physical modification of the Teflon-AF surface thus preserving the relevant properties (such as biocompatibility, inertness) of this material that render it valuable for regenerative medicine applications. As a consequence of a well-performed laminin pattern on Teflon-AF, I obtained an extremely selective cell adhesion, both among and along laminin stripes, with a control also on the cell number. This approach represents a viable route to the use Teflon-like or other highly hydrophobic materials in tissue engineering as a vector of cells in tissues (e.g. to deliver stem cells). The outcome is also a support for the controlled localization and orientation of the cells that is known to play a crucial role in the realization of complex tissues.

References

1. Saneinejad S, Shoichet MS (1998) J Biomed Mater Res 42:13–19
2. Pasche S, Voros J, Griesser HJ, Spencer ND, Textor M (2005) J Phys Chem B 109:17545–17552
3. Andersson AS, Glasmastar K, Sutherland D, Lidberg U, Kasemo B (2003) J Biomed Mater Res A 64:622–629
4. Langer R, Tirrell DA (2004) Nature 428:487–492
5. Wong EY, Thorne ML, Nikolov HN, Poepping TL, Holdsworth DW (2008) Ultrasound Med Biol 34:1846–1856
6. Scharnberg M, Zaporojtchenko V, Adelung R, Faupel F, Pannemann C, Diekmann T, Hilleringmann U (2007) Appl Phys Lett 90:013501
7. Pannemann C, Diekmann T, Hilleringmann U, Schurmann U, Scharnberg M, Zaporojtchenko V, Adelung R, Faupel F (2007) Mater Sci Poland 25:94–101
8. Spijkman MJ, Brondijk JJ, Geuns TCT, Smits ECP, Cramer T, Zerbetto F, Stoliar P, Biscarini F, Blom PWM, de Leeuw DM (2010) Adv Funct Mater 20:898–905
9. Makohliso SA, Giovangrandi L, Leonard D, Mathieu HJ, Ilegems M, Aebischer P (1998) Biosens Bioelectron 13:1227–1235
10. Ranieri JP, Bellamkonda R, Bekos EJ, Gardella JA Jr, Mathieu HJ, Ruiz L, Aebischer P (1994) Int J Dev Neurosci 12:725–735
11. Anamelechi CC, Truskey GA, Reichert WM (2005) Biomaterials 26:6887–6896
12. Cavallini M, Biscarini F (2003) Nano Lett 3:1269–1271
13. Timpl R, Rohde H, Robey PG, Rennard SI, Foidart JM, Martin GR (1979) J Biol Chem 254:9933–9937
14. Bustamante C, Keller D (1995) Phys Today 48:32–38

Chapter 4
Stamp-Assisted Multiscale Patterning of TiO_2 for Cell Growth Control

4.1 Introduction

Multiscale topographical features act on the interfacial forces guiding the cytoskeleton deformation and assembly and membrane receptor distribution thus governing cell adhesion and proliferation processes, cell morphology [1, 2] and gene expression [3]. A class of materials extremely promising for different applications, such as biomedical devices, scaffolds, catalysis and photocatalysis, is represented by bioceramics [4]. Among these, titanium dioxide (TiO_2) is a multifunctional biomaterial with long established application in bone replacement [5]. The formation of a native thin film of TiO_2 (3–6 nm) on titanium prosthesis is in responsible for the high biocompatibility of the implants [6, 7]. TiO_2 has a potential use also in new areas of regenerative medicine as tissue engineering for the fabrication of biocompatible 3D-scaffolds for bone tissue regeneration [8–10]. Moreover there is a soaring interest in the in vitro studies of neurodegenerative diseases for controlling the formation of neural networks both for tissue regeneration and cell-therapies [11, 12] and for transduction of cellular signals [13]. The interest around titanium dioxide is drawn by the possibility to employ TiO_2-based materials in implantable neuroprostheses and electro-stimulators due to their anti-inflammatory activity [14] and the possibility for TiO_2 to behave both as an electrical insulator and conductor [15, 16]. Recently the possibility to create a neural network of cerebral cortex neurons grown on anatase TiO_2 films with different electrical characteriztics and topographies has been demonstrated [17]. Despite these efforts, the investigation of the correlation between the environment local topography and neural cell growth requires a better control on the TiO_2 structure.

In this chapter, a statistical investigation of the adhesion and proliferation of SH-SY5Y neuroblastoma and 1321N1 astrocytoma cells on multiscale anatase TiO_2 patterned on glass by Micro-Molding in Capillary (MIMIC) technique is reported. Patterns have feature length scale from tens micrometers down to few hundreds nanometers and a controlled multiple length scale porosity (Fig. 4.1).

M. Bianchi, *Multiscale Fabrication of Functional Materials for Regenerative Medicine*, Springer Theses, DOI: 10.1007/978-3-642-22881-0_4,
© Springer-Verlag Berlin Heidelberg 2011

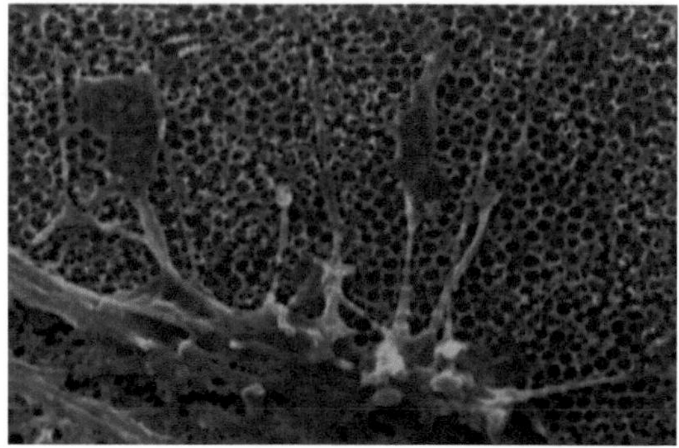

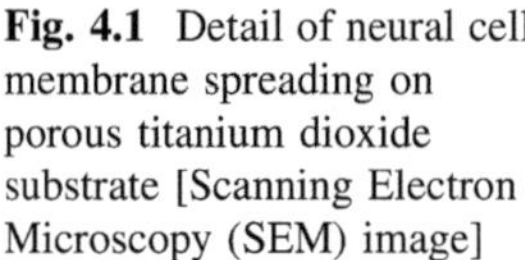

Fig. 4.1 Detail of neural cell membrane spreading on porous titanium dioxide substrate [Scanning Electron Microscopy (SEM) image]

I analyzed the dependence of cell density with respect to the porosity of the system, to the lateral confinement produced by the TiO$_2$ stripe width and to the glass background. The behaviour of the two neural cell lineages in response to the same topographical environment is also discussed.

4.2 Fabrication of Porous and Flat TiO$_2$ Patterns

Multiporous anatase TiO$_2$ patterns (MP) on glass substrate have been fabricated by MIMIC (Fig. 4.2). I chose the micrometric size of the polydimethylsiloxane (PDMS) stamp channels (Fig. 4.2a) in order to match integer multiples of the mean lateral size of a neural cell used in our experiments (20–25 μm), to ideally accommodate from one up to about three cells side-by-side on the same stripe. A drop of PS/TALH composite solution is deposited at the open end of the cavity of the PDMS stamp and the solution infills the channel upon Laplace pressure (Fig. 4.2b). During water evaporation at room temperature, the polystyrene beads (PS) pack into a hexagonal close structure with Bis(ammonium lactate)titanium dihydroxide (TALH) intercalated in the holes among beads (Fig. 4.2c, d). To obtain a regular multiporous array of TiO$_2$ in anatase crystalline phase, burning at the same time the PS and all the organic phase off, the sample is put into a furnace to be calcinated at 450°C in air (Fig. 4.2e). Flat patterns (FP) have been fabricated starting from TALH without PS, according to the same protocol used for MP.

Typical optical images of FP, PS/TALH patterns and MP are shown in Fig. 4.3a–c, respectively. Each pattern is made of continuous stripes stretching across a large area of the glass substrate. Figure 4.3d–i obtained by Atomic Force Microscopy (AFM) evidence the different morphology of flat (Fig. 4.3 left column) and porous TiO$_2$ stripes (Fig. 4.3 right column) whose 3D-multiporous structure is shown in Fig. 4.3i. TiO$_2$ stripes are a few hundred nanometers thick as evident from the line profiles in the insets. The FP morphology reveals small grains ca 10–15 nm diameter aggregated into a disordered film (Fig. 4.3g). Figure 4.3h shows the short range ordered pattern created by the PS beads. After

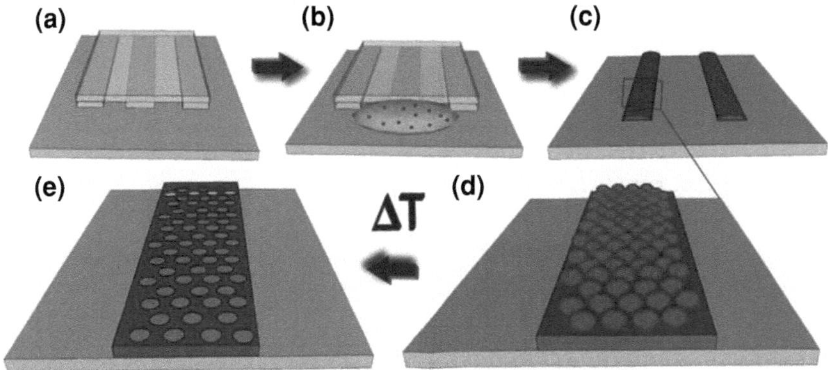

Fig. 4.2 MIMIC technique applied to the fabrication of multiporous anatase patterns. The PDMS stamp displaying micro-channels is put on the glass substrate (**a**); a drop of TALH/PS is deposited at the open end of the stamp and infills the channels (**b**); after solvent evaporation, stripes of TALH/PS are obtained (**c**), with the PS beads packed in according to a hexagonal arrangement (**d**); after thermal treatment at 450°C, PS beads are removed, the calcination process transforms the precursor TALH into anatase TiO₂ and the porous structure is obtained (**e**)

calcination, the mean diameter of the pores is 241 ± 5 nm. Since the PS diameter before calcinations was 340 ± 5 nm, the contraction of the volume due to the thermal treatment is -29%.

The Replica Molding (RMs) roughness is 0.5 ± 0.1 nm for FP, 73 ± 9 nm for MP, and 2.1 ± 0.3 nm for glass. The contact angles for the FP-film and glass do not significantly differ: $28 \pm 2°$ for the FP film and $25 \pm 2°$ for the glass slide. The MP-film is completely wetted by the water drop, indicating that the internal surface of the pores is highly hydrophilic despite of the large roughness value which opposes wetting [18].

4.3 Neural Cell Adhesion and Proliferation on TiO₂ Patterns

In Fig. 4.4 optical microscopy images show the typical distribution of both cell types used in this study on patterned stripes and their alignment after 48 h of incubation in standard condition.

I carried out a statistical analysis of adhesion and proliferation processes of astrocytoma 1321N1 and neuroblastoma SH-SY5Y cells on TiO₂ patterns starting from the set of optical images as in Fig. 4.4 by means of a semi-automatic approach (ImageJ software). Cell density on MP, FP and glass substrate have been extracted and reported in Fig. 4.5.

At three hours incubation, there is no statistical difference between cell density on glass and TiO₂ ($p > 0.05$). Evolution at later times reveals the sensitivity of the cells to the TiO₂ patterns. Both cell lines proliferate faster on TiO₂ stripes than on the bare glass (Fig. 4.5a, b). The trends observed in time can be fit with a linear regression.

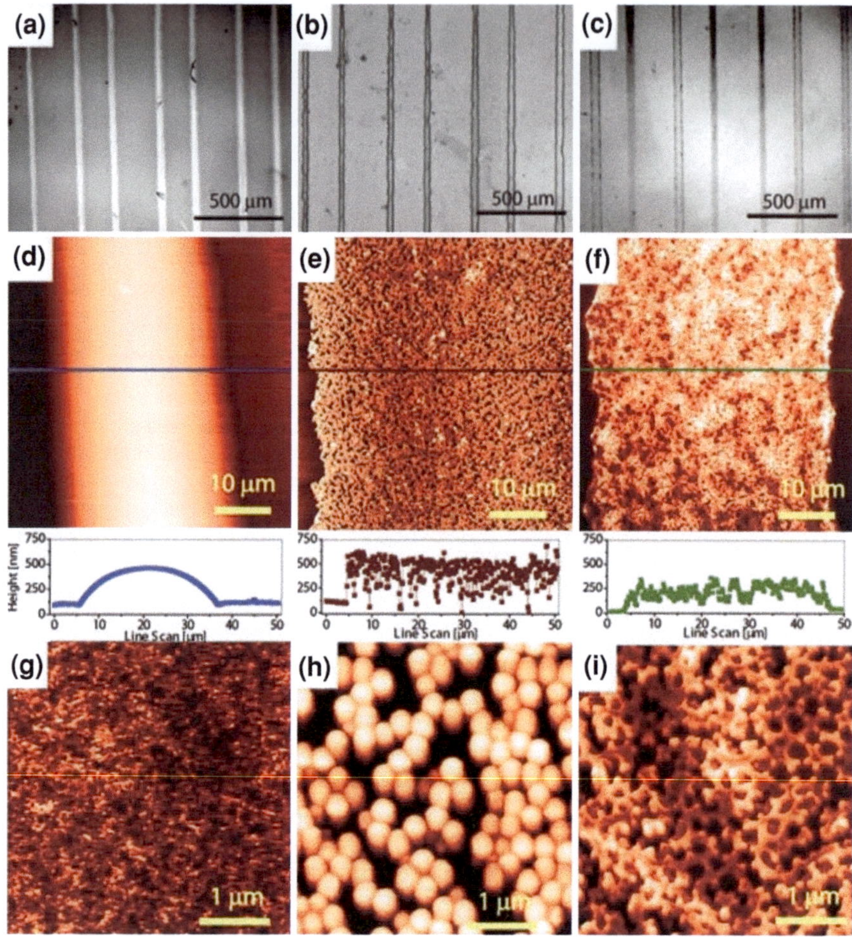

Fig. 4.3 Optical images (bright field) of FP (**a**) and MP before (**b**) and after (**c**) the thermal treatment. **d–f** shows AFM images (50 × 50 μm) of **a–c** stripes relatively, with line profiles. In AFM **g–i** images (4 × 4 μm), more details are showed: (**g**) the almost flat surface of FP, (**h**) the packing of the PS beads in the untreated TALH/PS stripe and (**i**) the multiporous structure of MP after thermal treatment

The proliferation rate of 1321N1 astrocytoma (R_{MP} = 1.15 ± 0.10 mm^{-2} h^{-1}; R_{FP} = 1.05 ± 0.27 mm^{-2} h^{-1}) is about 50% lower than that of SH-SY5Y neuro-blastoma cells (R_{MP} = 1.95 ± 0.27 mm^{-2} min^{-1}; R_{FP} = 2.19 ± 0.38 mm^{-2} min^{-1}). Within the estimated errors, no significant proliferation difference is observed between MP and FP within the same cell line (R_{MP}/R_{FP} = 1.09 for 1321N1 cells; R_{FP}/R_{MP} = 1.12 for SH-SY5Y cells). As a comparison, the evolution of the cell number density on glass is almost flat (R = −0.11 ± 0.15 mm^{-2} h^{-1} for 1321N1 cells) or slowly varying in time (R = 0.30 ± 0.19 mm^{-2} min^{-1} for SH-SY5Y cells). These data show a marked preference for TiO$_2$ surface, irrespectively of

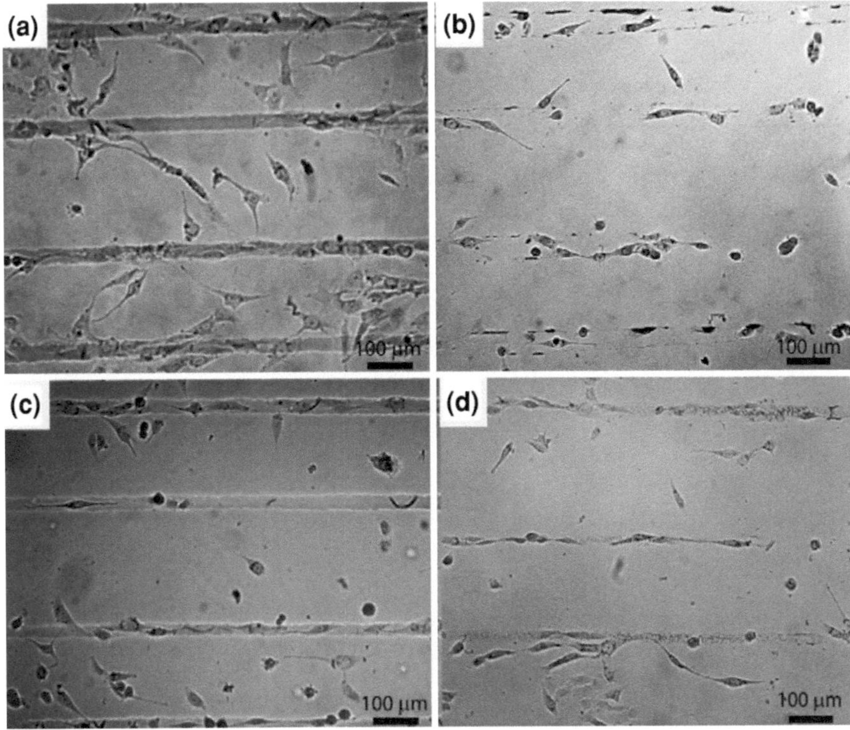

Fig. 4.4 Typical optical images (bright field) of astrocytoma 1321N1 (**a, b**) and neuroblastoma SH-SY5Y (**c, d**) cells on FP (**a, c**) and MP (**b, d**) at 48 h from seeding

its multiporous architecture, with respect to the glass background. Figure 4.5c and d shows the evolution of the ratio between the cell density on TiO₂ and on glass. Cell density ratio increases from about 2 times at 3 h to more than 4 times in 48 h then saturates (1231N1 cells) or decreases (SH-SY5Y cells). This can be reasonably ascribed to slowing down of proliferation on TiO₂ due to crowding and/or confluence on the stripes, accompanied by a subsequent increase of the density on glass. Tonazzini et al. showed that 1321N1 astrocytoma cells are able to adhere and proliferate on bare glass at higher rate than the one found in our experiments [19]. This evidence, along with our data, demonstrates that, in the presence of adjacent TiO₂ stripes, cell migration has occurred from glass to TiO₂. The driving force is likely the chemical gradient between titanium oxide surface and glass surface rather than the different morphology, as we do not observe dependence of proliferation rate on the multiporous architecture. It should be also noted that FP stripes have the same RMs and contact angle values as glass, thus reducing the differences between the two substrates only to the chemical nature (terminal groups, crystallinity). This fact should be remarked, as many papers report increased cell adhesion and proliferation by the introduction of some porosity. Our results seem to indicate some exception to this commonly accepted behaviour. Morphological features of both 1321N1 and

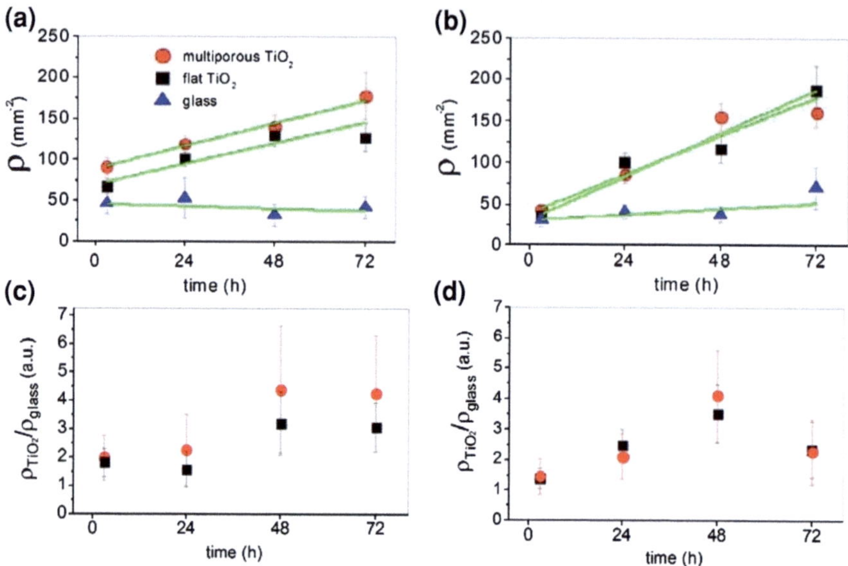

Fig. 4.5 Astrocytoma 1321N1 (**a**) and neuroblastoma SH-SY5Y (**b**) cell adhesion (3 h) and proliferation (24, 48 and 72 h) on MP, FP and glass. Statistical analysis: *filled circle* versus *filled square* p > 0.05; *filled circle* or *filled square* versus *filled triangle*: p < 0.01 (One-Way ANOVA analysis, Tukey's post test, Prism5, GraphPad Software). In (**c**) and (**d**), the cell density ratio between cell density on TiO$_2$ and on glass is reported both for 1321N1 and SH-SY5Y cell lineages, respectively

SH-SY5Y cells adhered to porous architecture of the MP are visualized by AFM and SEM (Fig. 4.6) analysis. Specifically, Fig. 4.6a and b shows details of the alignment of 1321N1 and SH-SY5Y cells on TiO$_2$ stripes, respectively. Figure 4.6c–f evidence numerous processes developing from both the cell phenotypes in order to adhere to the substrate. The development of peripheral lamellipodia and filopodia extensions and the spreading of the cells are well-recognized signs of a good cell anchoring to the substrate.

The biocompatibility of TiO$_2$ patterns has been also assessed. At the maximum incubation tested time, the cells display their typical morphology and appear green upon staining with the specific fluorescent dye fluorescein diacetate, indicating viable cells.

4.4 Neural Cell Density Dependence on Stripe Width

A finer analysis of the cell density versus stripe width has revealed new insights on how the cells sense their local environment. In Fig. 4.7, the time evolution of cell density in response to the lateral stripe width is shown. 1321N1 cell density (Fig. 4.7a, b) exhibits a fluctuation with two maxima both on FP and MP

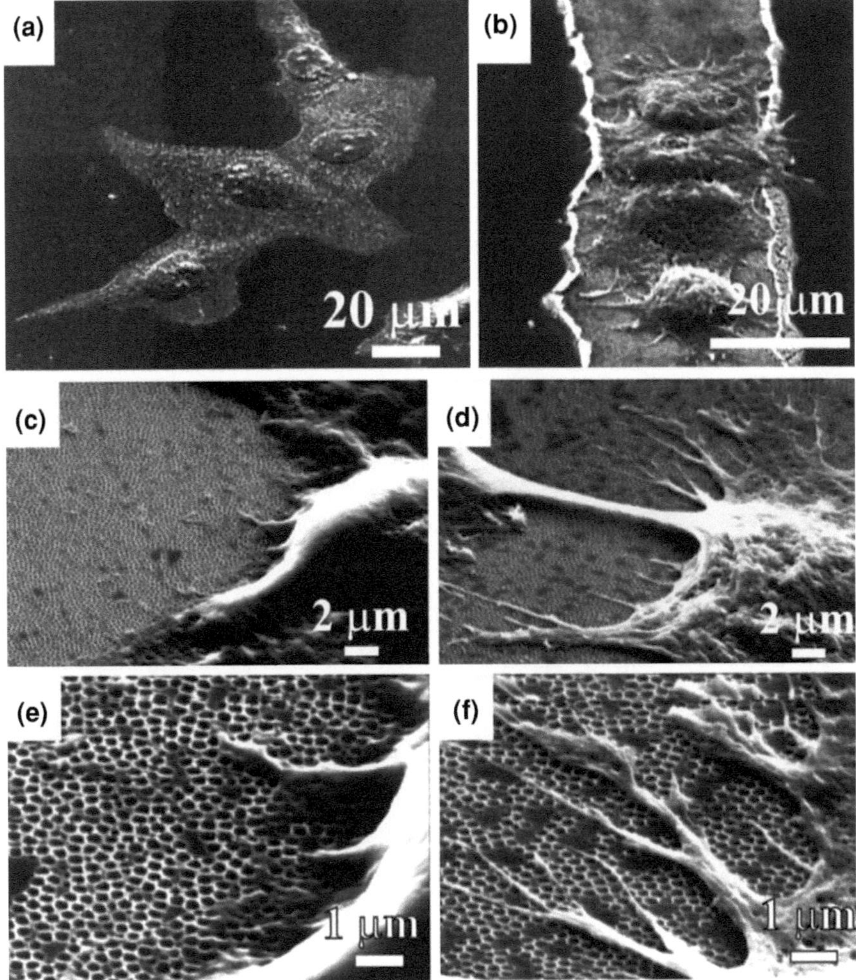

Fig. 4.6 SEM images of astrocytoma 1321N1 (**a**, **c**, **e**) and neuroblastoma SH-SY5Y (**b**, **d**, **f**) cell details on MP at 48 h. Figure **a** and **b** points out the cell alignment within a single stripe, while c–f shows details of the cell membrane spreading and protrusion development on the porous TiO_2 structure

corresponding to approximately one and 2–3 cell "diameters". The peaks on MP are more than 50% higher than the ones on FP.

This results suggest that astrocytoma cells distinguish both width and, to a minor extent, multiscale porosity. The latter does not appear clear from the coarse–grained density analysis reported in Fig. 4.3. On the contrary, the graphs of neuroblastoma cell proliferation show a single peak at the smallest width corresponding to the size of one cell. There is no difference in peak height versus porosity, suggesting that, along with data in Fig. 4.5, in our experimental

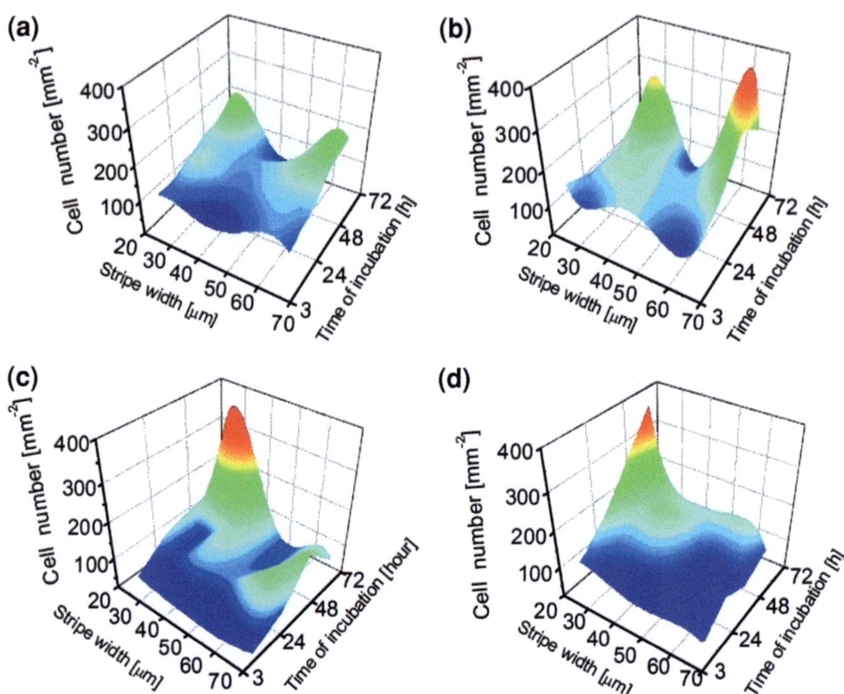

Fig. 4.7 1321N1 astrocytoma (**a, b**) and SH-SY5Y neuroblastoma (**c, d**) cell density 3D-plot with respect to the TiO$_2$ stripe width and proliferation time on FP (**a, c**) and MP (**b, d**). Data have been analyzed and smoothed with OriginPro 8.1 software

conditions neuroblastoma cells are not influenced by the multiscale porosity. The different behaviour of astrocytoma and neuroblastoma cells with respect to the lateral size of the stripes may be related to their different shape and cytoskeleton organization when growing on stripes. Astrocytoma cells are more elongated tooking up less room at 48 h, allowing the occupation of more than one cell per stripe; instead neuroblastoma cells exhibits a more neuron-like shape, occupying more surface area thus limiting the alignment to just one cell at a time on average. Although the data reported in Fig. 4.5 do not show significant differences in cell behaviour due to the presence of porosity, a slight preferential growth of astrocytoma cells on MP is detectable as observed in Fig. 4.7a and b.

A more accurate analysis of the cytoskeleton organization of astrocytoma on FP and MP has been performed. In Fig. 4.8 typical immunofluorescence images at 48 h are shown. Actin cytoskeleton was stained in red by TRITC-conjugated Phalloidin, focal adhesions in green by anti-vinculin and nuclei in blue by DAPI fluorescence. Figure 4.8a and b shows the elongated cell shape and the alignment on the TiO$_2$ stripe as well as the formation of several vinculin-containing focal adhesion points (Fig. 4.7c) on the multiscale porous architecture. However, from a qualitative analysis of several images as the ones reported in Fig. 4.8, it was no

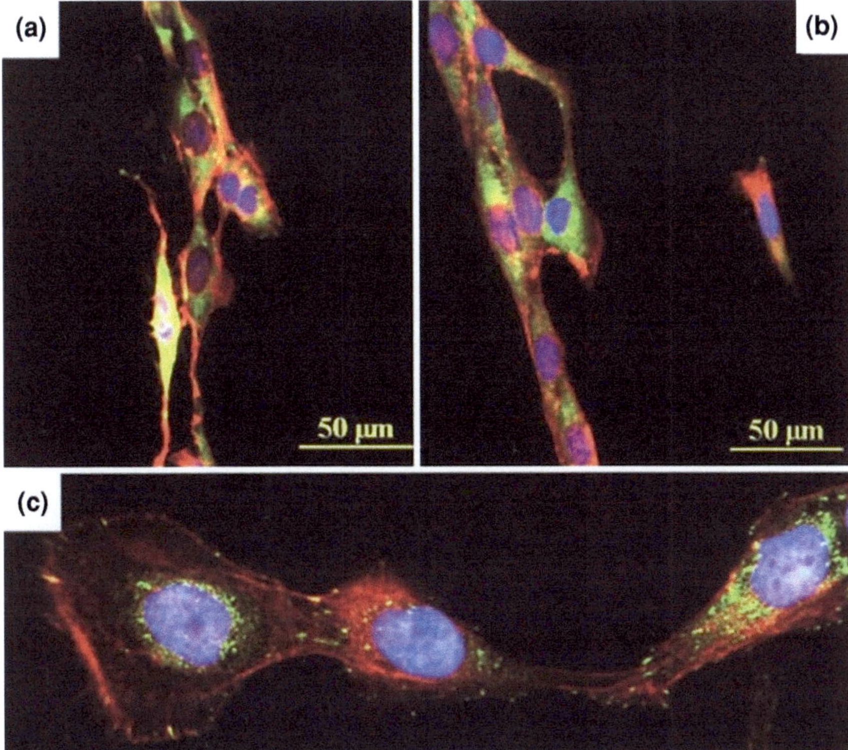

Fig. 4.8 Immunofluorescence staining optical images evidencing the alignment on human 1321N1 astrocytoma cells on FP (**a**) and MP (**b, c**). In (**c**) the cell focal adhesions are well visible

possible to detect appreciable differences both in the cytoskeleton and in focal point distribution due to the presence of the multiscale porosity. The analysis needed to clarify better this point are out of the thesis scope.

4.5 Materials and Methods

4.5.1 Colloidal Composite Preparation

TALH (50% in water, Aldrich) was used as received without any further purification. Monodisperse PS (2.13% vol in water) with an average size of 340 ± 5 nm were prepared by surfactant free emulsion polymerization of styrene as described elsewhere [20]. TALH/PS composite suspension was obtained by mixing TALH with PS (1:16 v/v) [21].

4.5.2 Titanium Dioxide Pattern Fabrication

MP were fabricated by MIMICs. The master was fabricated according to the following protocol. A cover glass slide (thickness $\sim$1 mm, Thermo Fisher Scientific Inc., MA, USA) was rinsed in piranha solution (H_2SO_4:H_2O_2, 3:1 v/v, 10 min, 100°C) and dried with N_2. A few micron-thick layer of a positive photoresist (AR-P 3210, Allresist GmbH, Strausberg, Germany) was deposited by spin coating on the glass slide (Laurell Technologies Corporation Spin Coater, 2000 rpm, 60 s). A pre-baking procedure (5′ at 70°C followed by 25′ at 90°C) was performed before the exposure to the UV light ($\lambda = 350$ nm, exposure time $= 10$ s, Mask Aligner Karl Suss MJB40) through a high contrast mask exhibiting 22 variable-width (from 15 to 75 μm) and variable-distance (165 and 230 μm) parallel lines (length $\sim$5 mm). The exposed photoresist on glass was developed in a 0.2 M KOH solution to dissolve exposed areas. A post-baking treatment was needed to consolidate the master (30 min at 110°C). Once we obtained the master, we prepared the molds (or stamps) by RM [22], a process which consists in casting PDMS on the master (PDMS elastomer; Sylgard 184 Down Corning, 1:10 curing agent:PDMS) and curing the elastomer in oven for 3 h at 95°C. After peel off, the PDMS stamp was put in contact with a glass slide (thickness $\sim$1 mm, Thermo Fisher Scientific Inc., MA, USA) cleaned before with piranha solution (H_2SO_4:H_2O_2, 3:1 v/v). A drop (5 μl) of the PS/TALH composite was deposited at the open end of the microchannels and the solution infilled, driven by capillary forces. After water evaporation at room temperature, the PDMS stamp was gently lifted off leaving a PS/TALH pattern. In order to transform the precursor into crystalline anatase TiO₂ [23] and eliminating the PS, samples were put into a furnace to be calcinated at 450°C in air. After water evaporation, samples were thermally processed in air up to 450°C with a thermal ramp of 1.77°C min^{-1} followed by a 2 h stationary step at 450°C. FP were realized at the same time by starting only from TALH, following the same fabrication and thermal conditions used for multiporous patterns.

4.5.3 Titanium Dioxide Pattern Characterization

TiO₂ patterns were characterized by optical microscopy with a Nikon microscope (Eclipse 80i, Nikon Corporation, Tokyo, Japan) in bright field. Higher magnification details of the morphology were imaged by both AFM and SEM. Samples were imaged by AFM (Smena, NT-MDT, Moscow, Russia) operating in semi-contact mode under ambient conditions with NSG10 cantilevers (NT-MDT, Moscow, Russia). AFM images were analyzed by using the software Image Analysis 2.2.0 (NT-MDT, Moscow, Russia). SEM images were collected with a SEM-FEG Hitachi S4000 ($V_{acc} = 20$ keV, $I = 10$μA) with out-of-plane angles of 0° and 45°. Samples for SEM analysis were metalized by sputtering a $\sim$5 nm gold layer on surface. The mean diameter of the macropores of MP was extracted by the statistical analysis of several SEM images (ImageJ free software, threshold tool); the RMs roughness of FP, MP and glass substrate was obtained by the analysis of different sized AFM images. The surface tension of TiO_2 patterns was correlated through water contact angle measurements at room temperature (Digidrop GBX contact angle model DS, France). For the measurement of the static contact angle with water droplets, thin TiO_2 films (2×2 cm, thickness $\sim$150 nm) were fabricated according to the same protocol used for patterns, this time without the stamp, with a Laurell Technologies Corporation spin coater (North Wales, PA, USA) following the experimental setup: $30''$, 100 acc., 1500 rpm.

4.5.4 Neural Cell Cultures

Human neuroblastoma SH-SY5Y cells (purchased from the European Collection of Cell Cultures, ECACC No. 94030304, Sigma Aldrich, St. Louis, MO, USA) were cultured in a 1:1 mixture of F-12 Nutrient Mixtures (Ham's F12) and EMEM (cell medium), supplemented with 15% Fetal Bovine Serum (FBS), 2 mM L-Glutamine, 1% Non Essential Amino Acids, Penicillin (100 U ml^{-1}) and Streptomycin (100 μg ml^{-1}, complete medium).

Human astrocytoma 1321N1 cells (from ECACC 86030402) were cultured in Dulbecco's Modified Eagle Medium: Nutrient Mixture F-12 (DMEM/F12), supplemented with 10% FBS, 2 mM L-Glutamine, Penicillin (100 U ml^{-1}), and Streptomycin (100 μg ml^{-1}) (complete medium). Each cell line was maintained in standard culture conditions ($37°$ C in a humidified atmosphere with 5% CO_2) and fed every 2–3 days. Cell culture materials were purchased from Cambrex Bio-Science (Verviers, Belgium), plastic wares from Sarstedt (Nümbrecht, Germany).

4.5.5 Neural Cell Adhesion and Proliferation Assays

At subconfluence, both neuroblastoma and astrocytoma cells were recovered by trypsinization step, suspended in their respective complete medium and seeded on both MP and FP at the final density of 3000 cm^{-2} and volume of 400 μl. The cells were allowed to proliferate under standard culture conditions for different times (from 3 to 72 h). To confine the cell culture on the area of interest, a circular PDMS pool ($\emptyset = 7.5$ mm, Sigma Aldrich, St. Louis, MO, USA) was placed on the sample around the patterned area ($\sim$25 mm^2). Experiments were done in duplicate and repeated at least five times. Before cell seeding, the samples were sterilized overnight under UV light. For adhesion experiments, optical images were taken after 3 h incubation time by means of a Nikon microscope (Eclipse-80i, Nikon Corporation, Tokyo, Japan) and cell density number estimated. Similarly, cellular proliferation was assessed by the analysis of optical images taken at 24, 48 and 72 h from seeding, after removal of cell medium to eliminate floating dead cells. Cell adhesion and proliferation on MP and FP were quantified in terms of cell density. A number n_i of images was collected for the i-th sample ($n_i = 20$–50, $n_{TOT} = 536$) at each time of incubation (3, 24, 48 and 72 h) over several significant pattern areas (1280 $\times$ 960 pixels). Adherent cells were estimated by counting them on both TiO$_2$ stripes and on glass with a semi-automatic approach (ImageJ free software) and the cell density (ρ [mm^{-2}]) values were extracted for MP, FP and glass. The proliferation rate (R [mm^{-2} h^{-1}]) was obtained from the slope of the graph ρ versus time. The cell proliferation ratio (ρ_{TiO2}/ρ_{glass} [a.u.]) was calculated by the ratio between cell density on MP or FP and on glass. To evaluate the cell behaviour with respect to the lateral dimension of TiO$_2$ stripes, the cell density was plotted versus stripe width, clustered in the following arbitrary intervals: 15–25, 26–35, 36–45, 46–55, 56–65 and 66–75 µm.

4.5.6 Neural Cell Viability Assay

To analyze neural cell viability, fluorescein diacetate (FDA, Sigma) staining was performed, as previously described [19]. At 48 and 72 h, cells were incubated for 5 min at 37°C with 75 µg ml^{-1} of dye solution and then rinsed with PBS 1X. After an additional washing step, the samples were observed under an inverted epifluorescence microscope attached to the optical stereomicroscope Olympus IX71 with SPM Biosolver (NTM-DT, Russia, Moscow). Living cells were stained in green, upon hydrolysis by intracellular esterases that yields fluorescein (fluorescent at excitation wavelengths of 490 nm). Cell shape analysis was performed with the software NIS Elements F 2.20 Nikon (Nikon Corporation, Tokyo, Japan).

4.5.7 Cytoskeleton and Focal Adhesion Staining by Immunofluorescence Assay

The formation of focal adhesion complexes and actin cytoskeleton of cells attached to MP and FP were assessed by using a specific kit (CHEMICON, Billerica, MA, USA) according to manufacturer's instructions. Cells were fixed with 4% paraformaldehyde, then permeabilized with 0.1% Triton X-100 for 2 min and simultaneously stained with TRITC-conjugated Phalloidin, anti-vinculin and DAPI in order to map the local orientation of actin filaments within cell, the focal contacts in cells and nuclei respectively. The samples were analyzed under an epifluorescence microscope endowed with NIKON excitation band DAPI-FITC-TRITC filter set and images were taken using a Nikon Instruments (Nikon, Tokyo, Japan) colour digital camera.

4.5.8 Cell Morphology Analysis by Combined SEM, AFM and Optical Microscopies

At different times of incubation, both astrocytoma and neuroblastoma cells grown on FP and MP were analyzed by optical, atomic force and scanning electron microscopy. For AFM analysis, the samples were fixed with 4% paraformaldehyde for 20 min at room temperature and then washed with PBS 1X. For SEM analysis (SEM-FEG Hitachi S4000, $V_{acc} = 20$ keV, I $= 10$ μA), cells were fixed with 2.5%, glutaraldheyde for 20 min at room temperature, washed with PBS 1X and then dehydrated in an increasing series of ethanol (25, 50, 75, 95 and 100% for 10 min each). Subsequently, a thin gold film coating (~ 5 nm) was sputtered before analysis.

4.5.9 Data Analysis

Graphic presentation and statistical analysis were performed using Graph-Pad Prism computer program (Graph Pad Software, version 5.0; San Diego, CA) and OriginPro 8.1 software (OriginLab Corporation, Northampton, MA, USA). All data were presented as mean $\pm$ SEM (standard error of the mean). The comparison of the data sets derived from the cell experiments was carried out statistically by One-Way ANOVA analysis (Tukey's HSD post test on pairwise treatments difference). P-value < 0.05 was considered statistically significant. P-value < 0.01 was considered highly statistically significant. The P-values were reported in the figure legends.

4.6 Conclusions

I have presented an unconventional and reliable approach for fabricating micro/nanostructured anatase TiO_2 patterns on glass for studying the response of different cell lines to a complex morphology of the surface.

This patterning technique yields sharp topography gradients and smoother chemical gradients due to the surface modulation of wetting/adsorption/hydrodynamic flow. Interestingly, the controlled multiscale features of the pattern can be opportunely tailored by choosing different colloidal monodisperse polystyrene beads size and stamp channel width. Neural cell adhesion and proliferation have been investigated in time as a function of cell phenotypes, presence of a multiporous architecture, stripe width. Both neural cell lines preferred to adhere on the TiO_2 stripe than on the glass background even the RMs roughness and the contact angle value were significantly similar. Proliferation rates on TiO_2 were larger than on glass, with neuroblastoma cell rate being twice that of astrocytoma ones. The latters were slightly more sensitive to the porous architecture than the former, even if there was no statistically significant discrepancy. Finally, the two cell lines exhibited a different behaviour versus stripe width, the former preferring widths matching integer multiple of their characteriztic size, and the latter with marked preference only for the width accommodating a single cell.

This work demonstrates that it is possible to exploit approaches based on non-conventional fabrication methods to perform systematic investigations on the relationship between topographical length scales and cell behaviour, with a conceivable future impact in the field of regenerative medicine.

References

1. Graeter SV, Huang JH, Perschmann N, Lopez-Garcia M, Kessler H, Ding JD, Spatz JP (2007) Nano Lett 7:1413–1418
2. Bystrenova E, Jelitai M, Tonazzini I, Lazar AN, Huth M, Stoliar P, Dionigi C, Cacace MG, Nickel B, Madarasz E, Biscarini F (2008) Adv Funct Mater 18:1751–1756
3. Dumont JE, Dremier S, Pirson I, Maenhaut C (2002) Am J Physiol Cell Physiol 283:C2–C28
4. Ohgushi H, Caplan AI (1999) J Biomed Mater Res 48:913–927
5. Feng B, Weng J, Yang BC, Qu SX, Zhang XD (2003) Biomaterials 24:4663–4670
6. Zhao XB, Liu XY, You J, Chen ZG, Ding CX (2008) Surf Coat Technol 202:3221–3226
7. Variola F, Vetrone F, Richert L, Jedrzejowski P, Yi JH, Zalzal S, Clair S, Sarkissian A, Perepichka DF, Wuest JD, Rosei F, Nanci A (2009) Small 5:996–1006
8. Moreno B, Carballo M, Jurado JR, Chinarro e (2009) Boletin De La Sociedad Espanola De Ceramica Y Vidrio 48:321–328
9. Wei J, Chen QZ, Stevens MM, Roether JA, Boccaccini AR (2008) Mater Sci Eng C-Biomimetic Supramol Syst 28:1–10
10. Tampieri A, Sandri M, Landi E, Pressato D, Francioli S, Quarto R, Martin I (2008) Biomaterials 29:3539–3546
11. Mahoney MJ, Anseth KS (2006) Biomaterials 27:2265–2274
12. Joers VL, Emborg ME (2010) Ilar J 51:24–41
13. Chang JC, Brewer GJ, Wheeler BC (2001) Biosens Bioelec 16:527–533
14. Suzuki R, Muyco J, McKittrick J, Frangos JA (2003) J Biomed Mater Res Part A 66A:396–402
15. Smith JR, Walsh FC, Clarke RL (1998) J Appl Electrochem 28:1021–1033
16. Ulbrich MH, Fromherz P (2005) Appl Phys A Mater Sci Process 81:887–891
17. Collazos-Castro JE, Cruz AM, Carballo-Vila M, Lira-Cantu M, Abad L, del Pino AP, Fraxedas J, Juan AS, Fonseca C, A. P. Pego, Casan-Pastor N (2009) Thin Solid Films 518:160–170
18. Koishi T, Yasuoka K, Fujikawa S, Ebisuzaki T, Zeng XC (2009) P Natl Acad Sci USA 106:8435–8440
19. Tonazzini I, Bystrenova E, Chelli B, Greco P, Stoliar P, Calo A, Lazar A, Borgatti F, D'Angelo P, Martini C, Biscarini F (2010) Biophys J 98:2804–2812
20. Dionigi C, Nozar P, Di Domenico D, Calestani g (2004) J Colloid Interf Sci 275:445–449
21. Dionigi C, Greco P, Ruani G, Cavallini M, Borgatti F, Biscarini F (2008) Chem Mater 20:7130–7135
22. Anderson JR, Chiu DT, Jackman RJ, Cherniavskaya O, McDonald JC, Wu HK, Whitesides SH, Whitesides GM (2000) Anal Chem 72:3158–3164
23. Ruani G, Ancora C, Corticelli F, Dionigi C, Rossi C (2008) Sol Energ Mat Sol C 92:537–542

Chapter 5
Control of Neural Cell Adhesion on 3D-SWCNT

5.1 Introduction

Neural cell adhesion, viz. the early stages following seeding on the substrate in vitro, involves shape modification leading to dendrite outgrowth, and appears important for guiding neuronal network formation. Electric fields have been extensively used to favor neurites (axons and dendrites) development and alignment in vitro [1–9]. Electric fields can also stimulate the repair of nerve injuries in both the peripheral and central nervous system in animal models [10, 11].

The early modifications of a cell, occurring when the electrical stimulation starts, have not been yet investigated. It can be expected that these are even more important as the cell has not enough time to explore and modify the environment [12]. On this basis it may be effective to combine topological/chemical cues with an electrical stimulus.

Carbon nanotubes (CNTs), which integrate mechanical, chemical (through functionalization) and outstanding electrical properties, are emerging as a potential substrate for promoting neuronal cell growth, viability and differentiation [13–19]. They have been shown to improve the signal transmission [20] and the responsiveness of neurons by the formation of tight contacts with cell membrane [21]. These studies were carried out on solution-deposited CNTs, and the effects of morphology, order or dimensionality of the network cannot be fully addressed. There is evidence that substrate dimensionality [22] and length scales [23, 24] are important factors in cell adhesion and growth. Patterned CNT networks would then be useful for investigating the effects induced by substrate length scales and dimensionality, maintaining at the same time the possibility to stimulate the cells with electric fields.

In this chapter, I explored the adhesion of SH-SY5Y neuroblastoma cells on 3D patterns of single-walled carbon nanotubes (SWCNT) semicapsules [25, 26] (Fig. 5.1). These patterns exhibit a controlled porosity at multiple length scales and electrical conductivity, both features being desirable for controlling the adhesion of

M. Bianchi, *Multiscale Fabrication of Functional Materials for Regenerative Medicine*, Springer Theses, DOI: 10.1007/978-3-642-22881-0_5,
© Springer-Verlag Berlin Heidelberg 2011

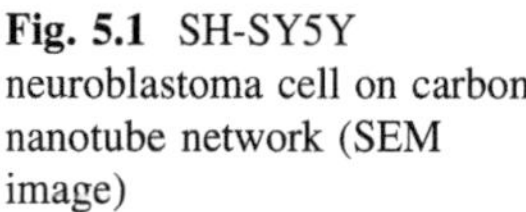

Fig. 5.1 SH-SY5Y
neuroblastoma cell on carbon
nanotube network (SEM
image)

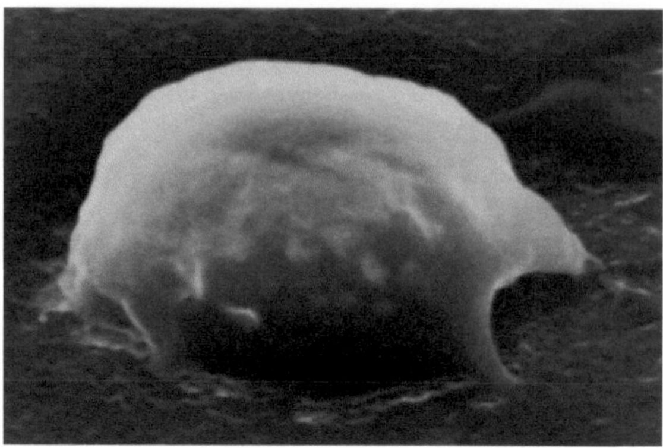

neuronal cells. This fact allow one also to assess the role of porosity with respect to solution-deposits of the same CNTs. The effect of the electric field applied across a stripe of a CNT array on the change of shape of SH-SY5Y cells has been also assessed.

5.2 Neural Cell Adhesion on SWCNT Pattern

5.2.1 Patterning of SWCNT by GAD

Sodium dodecyl sulfate (SDS), which is commonly used to suspend SWCNT, is demonstrated to be highly toxic for cells [27]. Therefore, to suspend the SWCNT I chose an aromatic amino, acid *O*-benzyl tyrosine (OBTY) which interacts with the SWCNT side wall through its aromatic group [28], as a bio-compatible suspending agent for SWCNT in the range $6.4 < \mathrm{pH} < 13$. A SWCNT/PS suspension was then prepared by mixing SWCNT suspension in OBTY with modisperse polystyrene beads (PS, Ø $\sim$ 400 nm).

Semicapsules made of SWCNT have been patterned on silicon wafer coated with native silicon oxide by grid assisted deposition (GAD) [29–32] of the SWCNT/PS colloidal composite. The GAD of the suspension on native silicon oxide layer is schematically depicted in Fig. 5.2a. A TEM Cu grid (pitch $\sim$ 500 μm, hole $\sim$ 420 μm) is gently placed on a drop of SWCNT/PS suspension and mechanically removed after water evaporation in ambient conditions, leaving the replica of the grid cavities on the substrate. The deposit is then rinsed with limonene to dissolve the PS. After rinsing, an array of SWCNT semicapsule squares is left on the substrate. SWCNT suspended with OBTY but in the absence of PS have been patterned by GAD to yield a pattern without semicapsule morphology. Figure 5.2b shows the SWCNT semicapsule pattern transferred by means of grid assisted deposition of SWCNT/PS suspension on the silicon wafer. From comparative

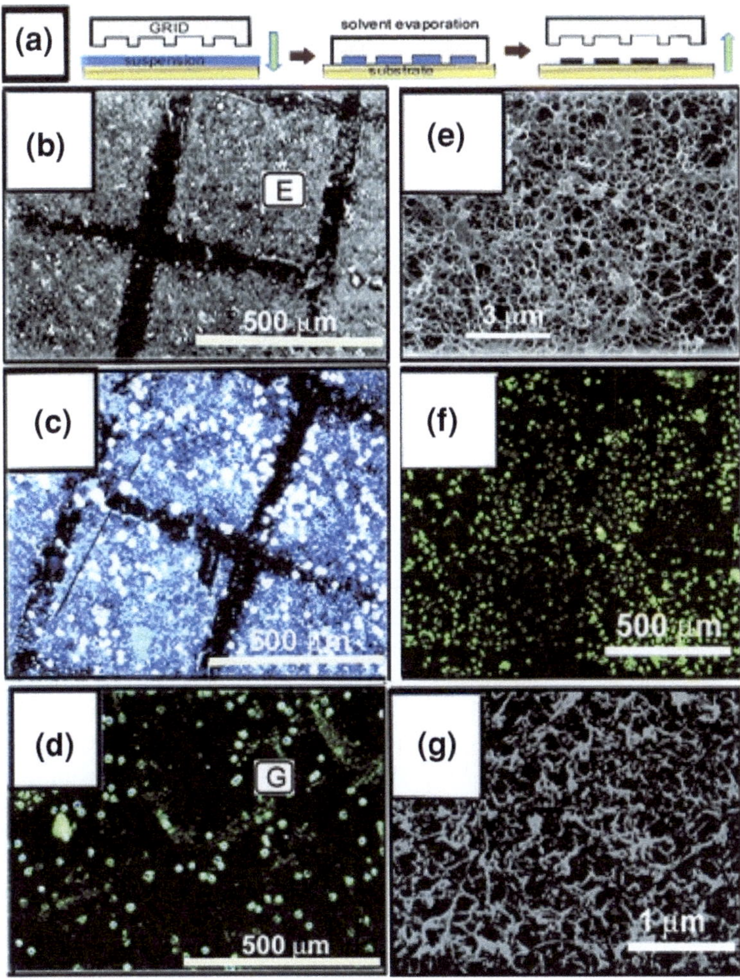

Fig. 5.2 Scheme of grid assisted deposition of the SWCNT suspension (**a**); SEM image of SWCNT semicapsules on silicon substrate (**b**); *dark field* images of SH-SY5Y cells at 30 min from seeding on patterned SWCNT (**c**) and disordered SWCNT (**d**); magnified SEM micrograph of the square regions consisting of semi-ordered SWCNT semicapsules (**e**); FDA assay on SH-SY5Y cells on patterned SWCNT 24 h after seeding (**f**); SEM image of the disordered SWCNT (**g**)

experiments with the same SWCNT suspension, the 400 nm sized PS have been turned out to represent the best suited diameter for the attachment of SWCNT bundles, that have to bend around the sphere and assemble in a close packing structure. The grid pitch is conserved in the deposited composite even after PS removal. SEM images show that the replica consists of a SWCNT semicapsule array with short range order (Fig. 5.2e). Since the square of the pattern has a 420 μm side, the coverage (i.e. the fraction of covered surface) of SWCNT is about 70%. The achievement of a proper replica either of the grid bars or of the grid holes strongly depends on the interplay of variables like (a) the grid size (equivalent experiments, not reported here, have been performed using a Cu grid with a pitch of 100 micron and yielded the replica of the grid bars leaving the centers of the squares empty), (b) the concentration of the suspension that determines the surface tension of the liquid drop, (c) the surface tension of the substrate and of the grid [33]. All these variables determine menisci of different curvatures. The composite suspension I used in the experiments have the proper composition to pin the meniscus in the cavity of the grid yielding a "solid square" deposit.

5.2.2 Cell Adhesion on SWCNT Array

SH-SY5Y cells have been suspended in cell medium at a final concentration of 2.5×10^4 cells cm^{-2} and seeded on the SWCNT networks with and without semicapsule architecture, for different times in order to test neuronal cell adhesion. The images show that SH-SY5Y cells adhere preferentially to the patterned SWCNT semicapsules (Fig. 5.2c). About 97% of the seeded cells (250/mm^2) adhere to the SWCNT pattern and only 3% of the cells have been found on the native silicon oxide surrounding the SWCNT squares. This corresponded to a density of 350 cells mm^{-2} on SWCNT and 27 cells mm^{-2} outside. Thus, the cells exhibit a marked preference for the SWCNT semicapsule pattern with respect to the native silicon oxide. For the PS free sample (Fig. 5.2d), the pattern is characterized by the inhomogeneous thickness of the SWCNT deposit in the squares. The presence of randomly assembled SWCNT is clearly shown by the SEM image in Fig. 5.2f. This inhomogeneity reasonably resulted from the comparatively larger capillary flow towards the periphery of SWCNT with respect to PS/SWCNT composite. As described in Ref. [25] this imply that for PS free SWCNTs the meniscus formed by the drop in the cavity is concave, and not convex. The enhanced contrast at the border is the result of dark field, viz. the impinging diffuse light was reflected largely by the edges. The cells are more randomly distributed on the pattern made of random SWCNT (Fig. 5.2d). The cell counting yielded 77% of the seeded cells (190 cells mm^{-2}) on SWCNTs, and 23% on native silicon oxide. This corresponded to a density of about 210 cells mm^{-2} on SWCNT, and 146 cells mm^{-2} outside. These densities are proportional to the areas occupied by SWCNTs, indicating that in the absence of the 3D pattern of the SWCNT the cells do not show any preference towards the SWCNT. SH-SY5Y cells remain viable both on the SWCNT semicapsule array (Fig. 5.2f) and on disordered SWCNT (Fig. 5.3) for at least 24 h as revealed by fluorescein diacetate (FDA) cell staining.

FDA tests have confirmed that SWCNT substrate maintains cell viability for a time period longer than the time needed for electrical stimulation.

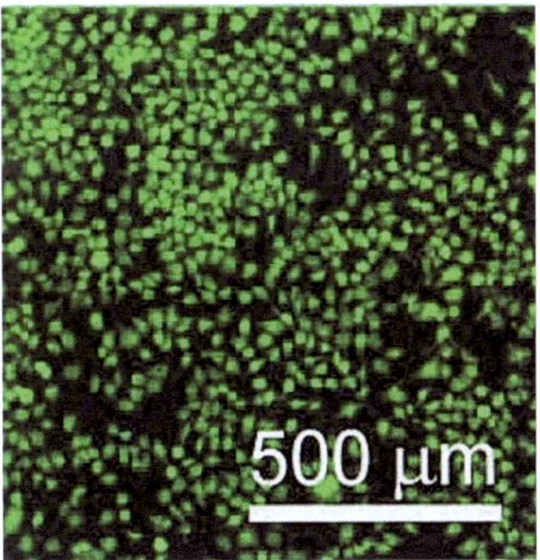

Fig. 5.3 FDA assay on SH-SY5Y cells on random SWCNT 24 h after seeding

In addition, experiments performed in the absence of SWCNT but only in the presence of the OBTY pattern, show that cells do not exhibit any preference for the OBTY pattern with respect to the silicon oxide substrate (Fig. 5.4).

Figure 5.4a is the optical photograph (bright field) of the OBTY pattern made on silicon oxide substrate with a copper grid with pitch size of 250 µm and hole size of 205 µm. Figure 5.4b represents the optical photograph (dark field) of SH-SY5Y cells after 30 min from seeding on the OBTY pattern. About 64% of the seeded cells (190/mm^2) has adhered on the pattern holes (silicon substrate), while only 36% of cells has been found on the OBTY pattern. This corresponds to a density of 208 cells/mm^2 on the pattern holes and only 136 cells/mm^2 on the OBTY pattern. Hence, any adhesion promoting effect due to the presence of OBTY during the experiment time can be excluded.

This combined evidences show that the 3D pattern is the main driving agency for preferential cell adhesion with respect to native silicon oxide, more effective than the random SWCNTs. Different topographical features affect cell adhesion, among them the multiscale porosity architecture can favour cell adhesion through the flow of nutrients and metabolites across adhering cells, and by promoting anchoring of focal contacts and/or actin–myosin stress fibers assemblies.

5.3 Electric Stimulation of Cells on SWCNT-Based Device

5.3.1 Test Bed Fabrication

I have fabricated a device test bed to pursuit electrical measurements on cells, consisting of planar junctions Pt/SWCNT semicapsules/Pt, following the procedure described in the "Materials and Method" part. The electrical characteriztics of the patterned Pt electrodes reveal an ohmic behaviour that has been consistently

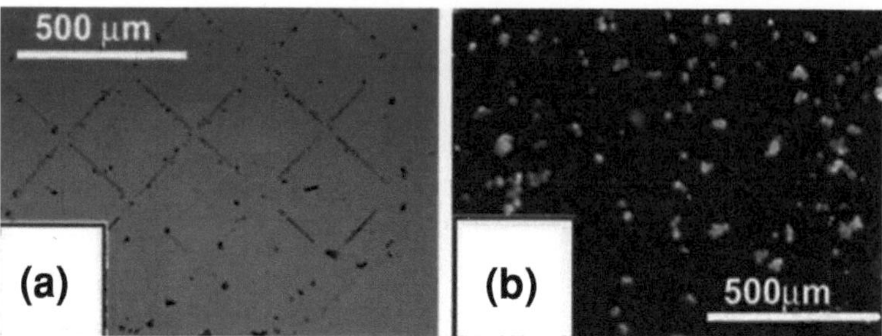

Fig. 5.4 Optical image (*bright field*) of the OBTY pattern made on SiO$_2$ substrate with a copper grid with pitch size of 250 µm (**a**); dark field image of SH-SY5Y cells at 30 min from seeding on the OBTY pattern (**b**)

Fig. 5.5 SEM images of **a** a SWCNT semicapsule stripe connecting the platinum electrodes. The inset shows the test bed pattern used in this work. **b** and **c** are magnified regions of the SWCNT semicapsule array in the stripes tilted at 45°

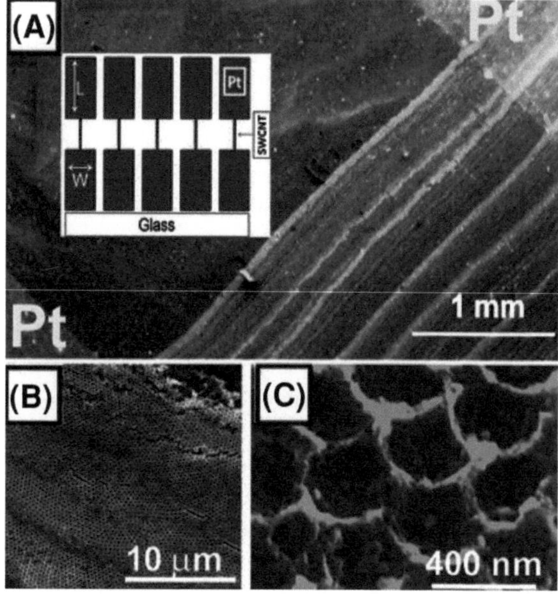

reproduced upon several voltage-sweep cycles. The sheet resistance of the Pt films is R$_s$ = 1 MΩ/□. SWCNT semicapsules have been integrated within the Pt planar electrodes by vertical deposition (glass slide immersed with a 55° tilt angle) [25] of the SWCNT/PS colloidal composite. PS have been removed by rinsing in limonene. A SEM image of a SWCNT sheet consisting of adjacent stripes formed by vertical deposition between two Pt contacts is shown in Fig. 5.5a and the test bed layout is shown in the inset. Magnified SEM images demonstrate that ordered arrays of SWCNT semicapsules are present within the stripes (Fig. 5.5b, c).

The current–voltage characteriztics of the patterned devices exhibits a good ohmic behaviour (Fig. 5.6, top). The order of magnitude of the sheet resistance R_s for all the stripes is 10^4–10^5 $\Omega/\square$. This parameter has been estimated considering a nominal aspect ratio $L\,W^{-1} = 2$ between each pair of Pt electrodes. The curves in Fig. 5.6 exhibit different slopes (vertical offset in the double-log plot). The spreading of the conductance, and hence of the sheet resistance, may be ascribed to fluctuations in the $L\,W^{-1}$ ratio from one stripe to another [25]. The stability and reproducibility of I–V characteriztics in water and in the cell culture medium with respect to air have been assessed (Fig. 5.4, bottom). There is a slight change of conductance moving from air to solutions, but the electrical responses of the SWCNT stripe in water and in medium are comparable. The stripes with larger extension of arrays of SWCNT semicapsules exhibit the lower sheet resistance and thus have been chosen for carrying out the electrical stimulation of seeded SH-SY5Y cells in culture medium.

5.3.2 Electric Stimulation

A voltage bias 15 min after SH-SY5Y cell seeding have been applied to the electrode pairs and maintained for 10 min. Electric field values corresponding to the applied bias voltages range from 0 V cm^{-1} up to 5 V cm^{-1} [1, 2]. Control experiments of SH-SY5Y adhesion on a glass slide and on a sheet of highly ordered pyrolytic graphite have been also carried out at $E = 0$ in the same experimental conditions. SEM images of fixed cells after electrical stimulation (Fig. 5.7a) show that cells adhere mainly to the SWCNT stripes with respect to the glass substrate. Therefore, the preferential cell adhesion observed in Fig. 5.2 is retained also on test beds/glass. SEM images display also that the cells under electric field (E) develop protrusions, even for 10 min application of electric field (Fig. 5.7b, d). The change of shape versus E is a clue to the enhancement of cell adhesion to the substrates.

I have first analyzed the cell shape on different substrates in the case of $E = 0$ V cm^{-1}. The resulting CSI values are shown in Fig. 5.8. No significant difference in CSI have been found between SWCNT and glass ($p > 0.05$), whereas there is a significant difference between SWCNT and graphite ($p < 0.05$). Thus, the cells exhibit a similar shape on SWCNT and glass in the very short time after seeding.

Then I have analyzed the influence of E. A shape modification in SH-SY5Y cells experiencing both $E = 1$ V cm^{-1} and $E = 3$ V cm^{-1} have been observed. The trend of the mean values of CSI and L versus E is reported in Fig. 5.8. The perimeter L increases versus E, and consistently CSI decreases, reflecting the enhancement of the "fluctuating rim morphology" in the presence of the electric field. The distributions of the perimeter values obtained from cells exposed to $E = 1$ V cm^{-1} and to $E = 3$ V cm^{-1} is significantly different from $E = 0$ V cm^{-1} ($p < 0.05$ and $p < 0.01$ respectively), and show 21 and 83% increase of their mean

Fig. 5.6 (*Top*) Current–voltage characteriztics of SWCNT semicapsule stripes in air and (*Bottom*) comparison of the characteriztics of stripe four in different media. The experimental data can be fit with a power law close to one in double log scale (*ohmic behaviour*)

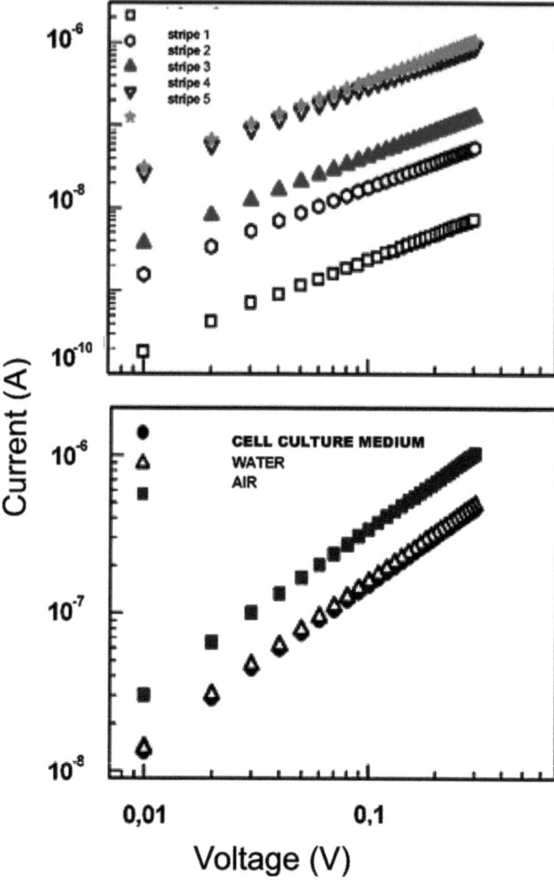

Fig. 5.7 **a–b** SEM images of SH-SY5Y cells immobilized with paraformaldehyde 4% showing preferential adhesion to SWCNT semicapsule stripes, under a 3 V cm^{-1} electric field (tilt angle of 45°). **c–d** SEM images of SH-SY5Y cells immobilized with paraformaldehyde 4% at **c** 0 V cm^{-1} and **d** 3 V cm^{-1}

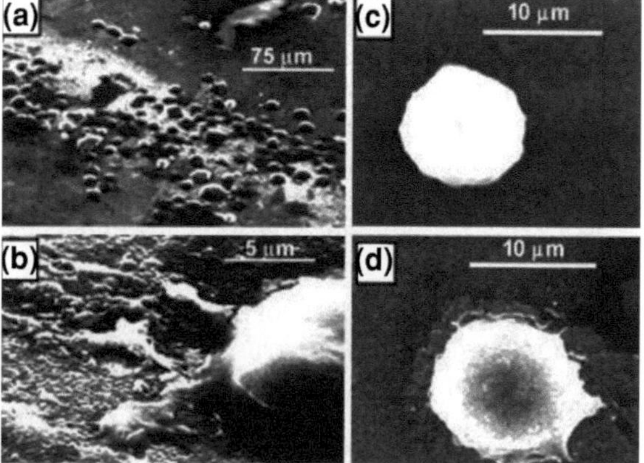

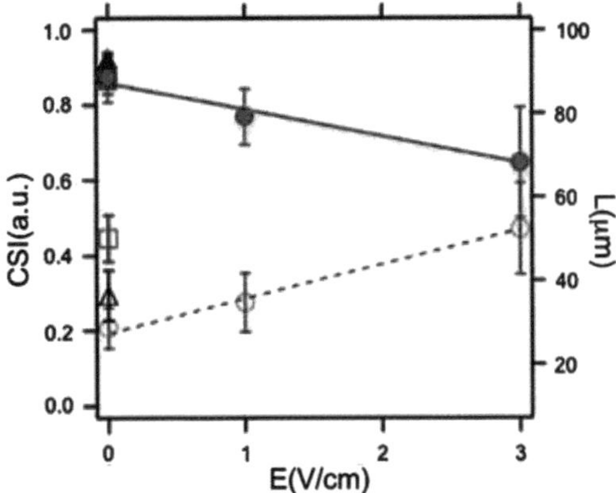

Fig. 5.8 Cell shape index CSI (*full markers, left axis*) and perimeter L (*empty markers, right axis*) of SH-SY5Y cells versus electric field. Circles refer to SWCNT semicapsules, squares to glass, triangles to HOPG

values with respect to $E = 0$ V cm^{-1}. The monotonically increasing trend of L versus E suggests a correlation between the electric field and the cell morphology, and hence cell adhesion. Similarly, CSI distributions exhibits a significant difference among the samples. At $E = 5$ V cm^{-1} a dramatic decrease in the number of cells adhering to the conductive stripes is observed with the apparent diameter decreasing by about 50% of the original one. This suggests that, at $E = 5$ V cm^{-1}, the adhesion of SH-SY5Y cells is inhibited by the electrical stimulus. It can not be excluded that the denaturation of the adhesion proteins secreted by the cells and adsorbed on the SWCNT has been induced by the electric field of 5 V cm^{-1} [34].

5.4 Materials and Methods

5.4.1 SWCNT Suspension in O-benzyl Tyrosine and SWCNT/PS Composite

O-benzyl tyrosine (OBTY) (Aldrich, No 14010) was used to suspend SWCNT. The suspension was obtained by sonicating 20 mg of SWCNT for 30 min in 50 ml of a [OBTY] $= 10^{-5}$ M aqueous solution. The solution was centrifuged and the supernatant resulted in an homogeneous suspension containing 0.3 mg ml^{-1} of SWCNT. SWCNT/PS suspension was prepared by mixing SWCNT suspension in OBTY with polystyrene beads (PS) in water to give a final colloidal composite containing 0.0013 g ml^{-1} of SWCNT and 0.0066 g ml^{-1} of monodisperse 400 nm diameter PS.

5.4.1.1 Patterning of SWCNT by GAD

Semicapsules made of SWCNT were patterned on silicon wafer coated with native silicon oxide by grid assisted deposition (GAD). The native oxide is a few nm thick and its water contact angle $\theta_c = (30 \pm 1)°$. TEM Cu grids (G50, SPI Supplies) with a pitch of 500 μm and hole size of 420 μm were used as template. The fabrication of SWCNT semicapsules was performed depositing 3 μl of the suspension on the substrate. The process is similar to the one illustrated in Ref. [25] (this time with the colloidal particles in place of a molecular or polymeric solute) and to those illustrated in Refs. [37, 38]. The Cu grid was gently placed on the drop and mechanically removed after water evaporation in ambient conditions, leaving the replica of the grid cavities on the substrate. The deposit was rinsed with limonene (p-mentha-1, 8-diene) for two minutes in order to dissolve the PS. After rinsing, an array of SWCNT semicapsule squares is left on the substrate. SWCNT suspended with OBTY but in the absence of PS were patterned by grid assisted deposition to yield a pattern without semicapsule morphology.

5.4.1.2 Cell Adhesion on SWCNT Arrays

Human SH-SY5Y neuroblastoma cells (purchased from the European Collection of Cell Cultures, ECACC No. 94030304) were cultured in a 1:1 mixture of F-12 Nutrient Mixtures (Ham12) and EMEM (cell medium), supplemented with 15% fetal bovine serum (FBS), 2 mM L-glutamine, 1% non essential amino acids, penicillin (100 U/mL) and streptomycin (100 μg mL^{-1}) (complete medium). Cells were maintained in standard culture conditions (37°C in a humidified atmosphere with 5% CO_2) and were fed every 2–3 days. SH-SY5Y cells were suspended in cell medium at a final concentration of 2.5 × 10^4 cells cm^{-2} and seeded on the SWCNT networks with and without semicapsule architecture, for different times in order to test neuronal cell adhesion. After cell incubation in standard culture conditions, the samples were rinsed with cell medium to remove non-adherent and dead cells before being observed under an optical microscope. Images were captured by an optical microscope (NIKON, Eclipse 80i) equipped with a Nikon digital camera. The cell viability was determined by fluorescein diacetate (FDA) cell staining [39].

5.4.1.3 Fabrication of Test Bed

A device test bed consisting of planar junctions Pt/SWCNT semicapsules/Pt was fabricated on a microscope glass slide to pursuit electrical measurements on cells. I first fabricated Pt electrodes (L = 3 mm, W = 1.5 mm) by MIMIC [29, 40, 41]. Briefly, a stamp made of PDMS with channels of 3 × 1.5 mm was placed on the glass surface. A dimethylformamide (DMF) solution of the Pt precursor $[NBu_4]_2[Pt_{15}(CO)_{30}]$ was poured at the inlets to infill the stamp channels by capillary pressure. This precursor has already been used for the fabrication of submicrometric conductive wires [38] and can be synthesized in water from commercial Pt salts [42]. After the complete evaporation of the solvent, the sample was annealed at a temperature of 150°C for 2 h to allow the complete decomposition of the carbonyl groups and the formation of Pt metal electrodes, with geometry defined by the stamp. The electrical characteristics of the patterned Pt electrodes revealed an ohmic behaviour that was consistently reproduced upon several voltage-sweep cycles. The sheet resistance [32] of the Pt films was $R_s = 1$ MΩ/. SWCNT semicapsules were integrated within the Pt planar electrodes by vertical deposition (glass slide immersed with a 55° tilt angle) [25] of the SWCNT/PS colloidal composite. PS were then removed by rinsing in limonene.

5.4.1.4 Electric stimulation

Electric stimulation of SH-SY5Y cells was performed in a closed home-made culture chamber, supplied with electric feedthroughs and equilibrated with the cell culture environment. Slabs of PDMS (2×2 mm, thickness ~ 1 mm), with a pool carved into them to contain the solution, were aligned on the electric device to confine the cell culture to the semicapsule stripes and to avoid contact with the electrodes. SH-SY5Y cells at a concentration of 2.5×10^4 cell cm^{-2} were seeded on the test bed from 10 μl of culture medium. A voltage bias was applied to the electrode pairs 15 min after seeding and maintained for 10 min. Electric field values corresponding to the applied bias voltages ranged from 0 V cm^{-1} up to 5 V cm^{-1} [1, 2]. Control experiments of SH-SY5Y adhesion on a glass slide and on a sheet of highly ordered pyrolytic graphite (HOPG supplied by NT-MDT), at $E = 0$ in the same experimental conditions were also carried out. After the electrical measurements, cells were rinsed with fresh culture medium to eliminate non-adhering and dead cells, then fixed with 4% paraformaldehyde and washed with phosphate-buffered saline (PBS).

5.4.1.5 Statistical Analysis

All data from cell experiments are reported as mean ± standard deviation (SD). The comparison of the statistical data sets obtained at different fields was carried out by One-Way ANOVA (Tukey's HSD test on pairwise treatments difference with IgorPro 4.07 Carbon, Wavemetrics, Lake Oswego, OR, USA). Statistical significance refers to results where $P < 0.05$ was obtained.

The shape modification of SH-SY5Y neuronal cells was evaluated following the model reported by Shen et al. [43]. After a few minutes of electrical stimulation. The projection area (S) and perimeter (L) of SH-SY5Y cells (including protrusions) were outlined by thresholding (Image SXM ver. 187-1c, based on NIH-Image 1.62, National Institute of Health, Bethesda MD). The Cell Shape Index (CSI) $= 4\pi S\ L^{-2}$ was extracted for each cell in the image and averaged. An elongated morphology corresponds to a CSI close to 0, whereas a circular shape yields CSI ~ 1 [43, 44].

5.5 Conclusions

In this chapter I demonstrated that a novel conductive substrate made of SWCNT patterned into regular semicapsules at a submicrometric scale can be exploited as a functional substrate for modulating the behaviour of SH-SY5Y neuroblastoma cells from preferential adhesion to non-adhesion. The dynamic control of cell adhesion onto a substrate during cell cultivation is an important issue for various biological and medical applications like tissue engineering. Neuroblastoma cells have been demonstrated to preferentially adhere to the SWCNT template. The latter have both structural and electrical functionalities. The well defined 3D porosity guarantees suitable conditions for cell viability and this is a desirable feature with respect to patterns fabricated by planar technology on silicon [35] or thin films of organic semiconductors [24]. On the other hand, I expect that the 3D-SWCNT features are comparable to those obtained by the 3D self-assembling of PS/organic semiconductor [36]. Moreover, the bio-functionalisation of SWCNT should also be explored for its possible use in biology. The conductance of the SWCNT network has allowed to perform a stimulation of the adhering cells by a DC electric field, inducing morphological changes in the cells towards a "fluctuating rim morphology" which preludes to the development of protrusions. I have found that it is possible to control cell adhesion by the electric field and that a maximum electric field value for the physiological functionality and expression of neuronal cells is around 5 V cm^{-1}. Only electric fields below 5 V cm^{-1} are suitable for promoting the adhesion of cells. In conclusion this part of the work highlights the importance of fine variations in voltage/current in the proximity of the cell focal points, which likely have implications in the signalling and the genotypical responses of the cells.

References

1. Patel NB, Poo M-M (1982) J Neurosci 2:483–496
2. Patel NB, Poo M-M (1984) J Neurosci 4:2939–2947
3. Al-Majed AA, Neumann MC, Brushart TM, Gordon T (2000) J Neurosci 20:2602–2608
4. McCaig CD, Zhao M (1997) BioEssays 19:819–826
5. Erskine L, McCaig CD (1995) Dev Biol 171:330–339
6. Greenebaum B, Sutton CH, Vadula MS, Battocletti JH, Swiontek T, DeKeyser J, Sisken BF (1996) Bioelectromagnetics 17:293–302
7. McCaig CD, Sangster L, Stewart R (2000) Dev Dyn 217:299–308
8. Zhang Y, Ding J, Duan W, Fan W (2005) Bioelectromagnetics 26:406–411
9. Liopo A, Stewart MP, Hudson J, Tour JM, Pappas T (2006) J Nanosci Nanotechnol 6:1365–1374
10. Sisken BF, Kanje M, Lundorg G, Herbst E, Kurtz W (1989) Brain Res 485:309–316
11. Borgens RB, Toombs JP, Breur G, Widmer WR, Waters D, Harbath AM, March P, Adams LG (1999) J Neurotrauma 16:639–657
12. Voldman J (2006) Annu Rev Biomed Eng 8:425–454
13. Mattson M, Haddon R, Rao AM (2000) J Mol Neurosci 14:175–182
14. Hu H, Ni Y, Montana V, Haddon RC, Parpura V (2004) Nano Lett 4:507–511

15. Hu H, Ni Y, Mandal SK, Montana V, Zhao B, Haddon RC, Parpura V (2005) J Phys Chem B Lett 109:4285–4289
16. Harrison S, Atala A (2007) Biomaterials 28:344–353
17. Zhang X, Prasad S, Niyogi S, Morgan A, Ozkan M, Ozkan CS (2005) Sens Actuators B 106:843–850
18. Lovat V, Pantarotto D, Lagostena L, Cacciari B, Grandolfo M, Righi M, Spalluto G, Prato M, Ballerini L (2005) Nano Lett 5:1107–1110
19. Gheith MK, Pappas T, Liopo V, Sinani VA, Shim BS, Motamedi M, Kotov NA (2006) Adv Mater 18:2975–2979
20. Mazzatenta A, Giugliano M, Campidelli S, Gambazzi L, Businaro L, Markramm H, Prato M, Ballerini L (2007) J Neurosci 27:6931–6936
21. Cellot G, Cilia E, Cipollone S, Rancic V, Sucapane A, Giordani S, Gambazzi L, Markram H, Grandolfo M, Scaini D, Gelain F, Casalis L, Prato M, Giuliano M, Ballerini L (2009) Nat Nanotechnol 4:126–133
22. Galvan-Garcia P, Keefer EW, Yang F, Zhang M, Fang S, Zakhidov AA, Baughman RH, Romero MI (2007) J Biomater Sci Polym Ed 18:1245–1261
23. Graeter SV, Huang J, Perschmann N, Lopez-Garcia M, Kessler H, Ding J, Spatz J (2007) Nanoletters 7:1413–1418
24. Bystrenova E, Jelitai M, Tonazzini I, Lazar AN, Huth M, Stoliar P, Dionigi C, Cacace MG, Nickel B, Madarasz E, Biscarini F (2008) Adv Funct Mater 18:1751–1756
25. Dionigi C, Stoliar P, Ruani G, Quiroga SD, Facchini M, Biscarini F (2007) J Mater Chem 17:3681–3686
26. Furtado CA, Kim UJ, Gutierrez H, Pan L, Dickey EC, Eklund PC (2004) J Am Chem Soc 126:6095
27. Hrenovic J, Ivankovic T (2007) Cent Eur J Biol 2:405–414
28. Chen RJ, Zhang Y, Wang D, Dai H (2001) J Am Chem Soc 123:3838–3839
29. Zhao X, Xia Y, Whitesides GM (1997) J Mater Chem 7:1069–1074
30. Cavallini M, Albonetti C, Biscarini F (2009) Adv Mater 21:1043–1053
31. Massi M, Cavallini M, Biscarini F (2009) Surf Sci 603:503–506
32. Massi M, Cavallini M, Stagni S, Palazzi A, Biscarini F (2003) Mater Sci Eng C Biomimetic Supramol Syst 23:923–925
33. Cavallini M, Biscarini F, Massi M, Morales AF, Leigh DA, Zerbetto F (2002) Nano Lett 2:635–639
34. Mendes PM (2008) Chem Soc Rev 37:2512–2529
35. Patolsky F, Timko BP, Yu GH, Fang Y, Greytak AB, Lieber CM (2006) Science 313:1100–1104
36. Dionigi C, Stoliar P, Porzio W, Destri S, Cavallini M, Bilotti I, Brillante A, Biscarini F (2007) Langmuir 23:2030–2036
37. Celio H, Barton E, Stevenson KJ (2006) Langmuir 22:11426–11435
38. Valentini L, Taticchi A, Marrocchi A, Kenny JM (2008) J Mater Chem 18:484–488
39. Low SP, Williams KA, Canham LT, Voelcker NH (2006) Biomaterials 27:4538–4546
40. Cavallini M, Serban DA, Greco P, Melinte S, Vlad A, Dutu CA, Zacchini S, Iapalucci MC, Biscarini F (2009) Small 5:1117–1122
41. Greco P, Cavallini M, Stoliar P, Quiroga SD, Dutta S, Zacchini S, Iapalucci MC, Morandi V, Milita S, Merli PG, Biscarini F (2008) J Am Chem Soc 130:1177–1182
42. Femoni C, Kaswalder F, Iapalucci MC, Longoni G, Mehlstaubl M, Zacchini S (2005) Chem Commun 46:5769–5771
43. Shen JY, Chan-Park MB, He B, Zhu AP, Zhu X, Beuerman RW, Yang EB, Chen W, Chan V (2006) Tissue Eng 12:2229–2240
44. Thakar RG, Ho F, Huang NF, Liepmann D, Li S (2003) Biochem Biophys Res Commun 307:883–890

Chapter 6
Lithographically Controlled Etching

6.1 Introduction

In the previous chapters it has been showed the unconventional fabrication of versatile systems useful for investigating the cell response to a particular environment signal, being either chemical, topographical or electrical. These systems can engender interest in the fields of regenerative medicine and bio-sensors, because of the possibility to address signals in a spatial controlled way [1, 2]. This is needed for enabling functionalities and connect nanomaterials in the device structures to the outside world [3].

Usually, in order to define the size, position and distances of nanostructures on a nanostructured surface, it is first necessary to fabricate the structured substrate, then to direct the assembly of the nano-objects by a second process. Either the functional materials or the substrates can be structured by several methods yielding nanometric resolution and important advantages for specific problems. However, none of the current techniques can be applied for the simultaneous patterning of both substrate and active material, in one step.

Since many devices and sensors that are based on organic and hybrid materials exhibit their best performances when fabricated on conventional, inorganic (Si/SiO$_2$), substrates [4], the use of a single-step method able to pattern a similar substrate and deposit/self-assembling addressable functional nanomaterials is highly desirable.

Here I propose an original process, termed Lithographically Controlled Etching (LCE), which yields the micro/nanostructuration of the substrate together with the spatially controlled deposition of a solute (molecules, supramolecular aggregates or nanoparticles) in a single step (Fig. 6.1). Thus LCE integrates in the same process both top–down and bottom–up methodologies [5], overcoming the limits of each approach and exploiting their advantages. I demonstrate that LCE is suitable for large area patterning and is sustainable because of its simplicity and reliability, yielding nanometer sized structures in a few minutes and in a single step. Since the process

M. Bianchi, *Multiscale Fabrication of Functional Materials for Regenerative Medicine*, Springer Theses, DOI: 10.1007/978-3-642-22881-0_6,
© Springer-Verlag Berlin Heidelberg 2011

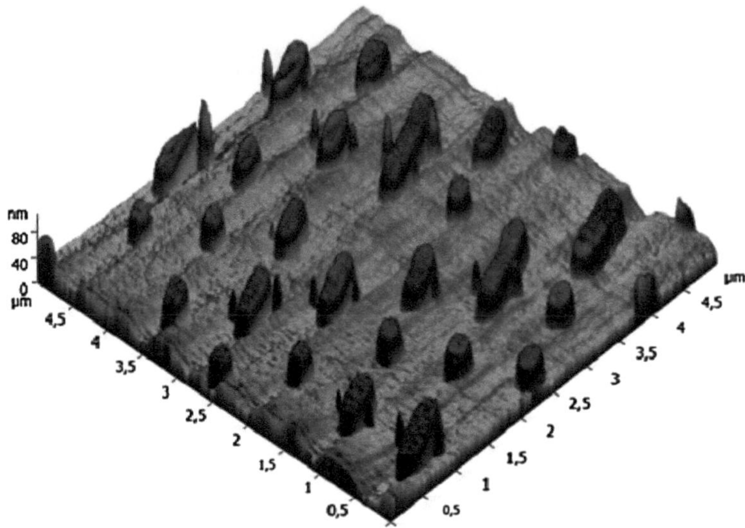

Fig. 6.1 Localized deposition of gold nanoparticles next to silicon oxide nanostructures (3D AFM image)

does not rely on specific interactions between the solution and the surface, LCE is of general application to a large variety of substrates and soluble materials.

6.2 Surface Patterning by Lithographically Controlled Etching

In LCE basic procedure (Fig. 6.2), a drop of etching solution (ES) is spread on a surface (a). When an elastomeric stamp is placed on the drop (b), the stamp protrusions come into contact with the surface thus protecting the surface underneath and giving rise to a thin layer cell in the recesses of a stamp motif (c). The ES reacts with the surface only under the stamp recesses and after exhaustion of the process, a patterned surface is obtained as a negative replica of the elastomeric stamp, (d).

The initial spreading of the drop on the surface (few seconds before placing the stamp) gives rise to a homogeneous distribution of the etching solution, and starts the reaction on the surface without any spatial confinement. This procedure can be applied only to mild ES, in order to avoid uncontrolled etching processes before stamp application. ES with etch rate <25 nm/min are suitable for LCE.

Despite of being a relatively weak acid (pKa = 3.15), HF exhibits a high etching power with respect to various oxides [6].

It is known that a hydrofluoridric acid (HF) aqueous solution etches isotropically (viz. with the same etch rate along each direction) a silicon oxide (SiO_2) surface [7].

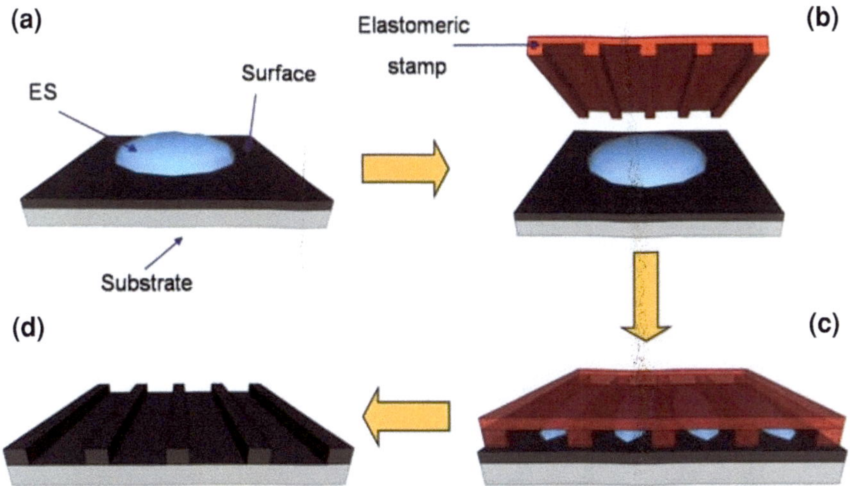

Fig. 6.2 Lithographically controlled etching general scheme

Table 6.1 Etch rates of thermal silicon oxide in several dilutions of HF

Etchant	Thermal oxide etch rate (nm/min)
Conc. HF(49%)	2,300
1:10(HF:H$_2$O)	23
1:25(HF:H$_2$O)	9.7
1:100(HF:H$_2$O)	2.3

This is a key step in many industrial fabrication techniques to remove the protective SiO_2 layer on Si wafers to be processed. The etch rate for a thermal SiO_2 layer grown on on a Si wafer and whose thickness is at least 200 nm, using different HF dilutions at room temperature (T $\approx$ 20°C) is reported in Table 6.1 [8].

The use of thermal SiO_2 as substrate and HF as ES is particularly efficient because the etch rate is nearly linear with HF concentrations in the 1:10 to 1:100 range, as can be inferred from Table 6.1. Thus one can calculate the right HF concentration in order to obtain the desired feature sizes at a given temperature, time, solution recirculation and stamp feature size [9].

A desirable feature of such etching process (Eq. 6.1) is that only the volatile compound SiF_4 and water are formed and no solid residues are produced during the process:

$$SiO_2(s) + 4HF\ (aq) \rightarrow SiF_4(g) + 2H_2O\ (l) \tag{6.1}$$

Figure 6.3 shows examples of LCE application on thermal SiO_2 wafer (oxide thickness $\sim$200 nm) for the fabrication of different micrometric and sub-micrometric patterns. They have been obtained starting from PDMS stamps patterned with: circular holes (Fig. 6.3b, pitch $\sim$10 µm, Ø $\sim$5 µm), linear structures

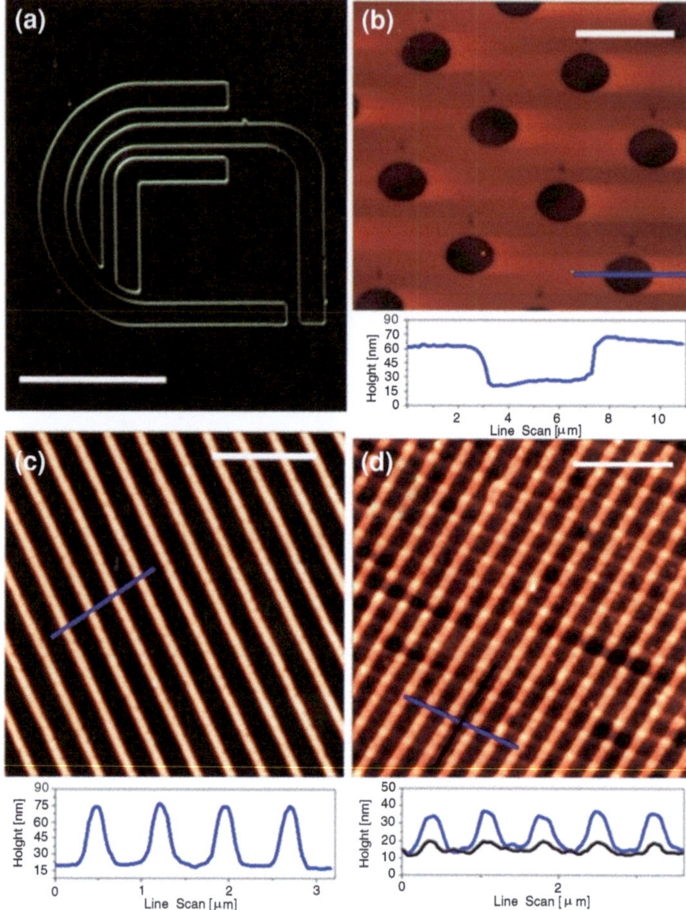

Fig. 6.3 Examples of LCE application. **a** optical micrograph (dark field) of CNR logo (ruler bar is 50 μm, Z scale 0–30 nm). **b** Regular distribution of micrometric holes fabricated on SiO_2/Si surface; (AFM image, bar 10 μm). **c** Parallel SiO_2 wires (AFM image, bar 3 μm). **d** Crossbars fabricated by iterating LCE process on sample **c** with the stamp rotated by 90° (AFM image, bar 3 μm)

(Fig. 6.3c, pitch ∼710 nm, FWHM ∼300 nm) and more complex drawings as the CNR logo (Fig. 6.3a, approximate printed size ∼0.1 × 0.1 mm,). As ES, a 0.4% (v/v) HF water solution was used.

The square feature in Fig. 6.3d demonstrates the possibility to iterate the LCE process. The pattern is obtained starting from parallel wires as in Fig. 6.3c (first LCE step), by performing a second LCE step with the new PDMS stamp rotated of about 90° with respect to the first one. A 3D elaboration of an AFM image showing a crossbar sample is reported in Fig. 6.4.

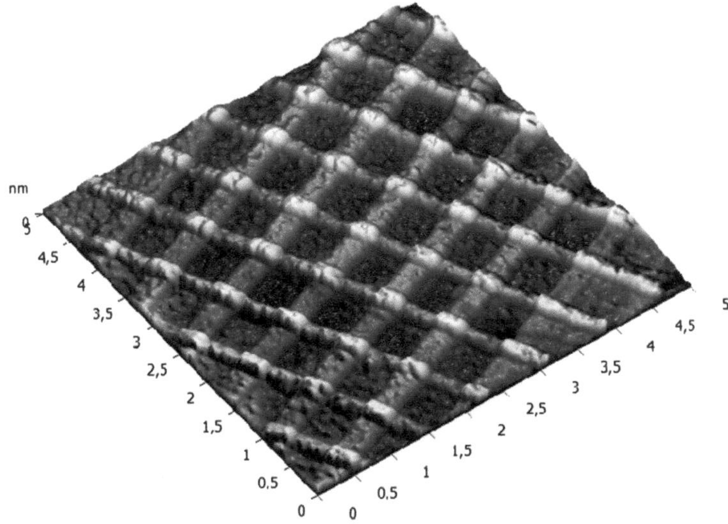

Fig. 6.4 LCE iteration on pattern with SiO_2 stripes by rotating the stamp of 90°

The larger SiO_2 stripes in Fig. 6.4 are obtained after the second LCE step. They present the same full-width at half maximum (FWHM ~300 nm) of the stripes in Fig. 6.3c (line profile). The thinner stripes derive from the first etching step, so they have been narrowed during the second LCE step (FWHM ~180 nm). The white areas where perpendicular stripes cross, correspond to the areas protected by the stamp in the second LCE application: that is why they are the highest feature in the image, whereas the other areas have been lowered as a consequence of the isotropic etching.

Figures 6.3d and 6.4 shows the potentiality of LCE for the achievement of relatively complex structures by a combination of multiple etching steps on the same sample.

LCE technique can be scaled up to the macroscopic scale (Fig. 6.5).

The optical image in Fig. 6.5a has been acquired with a digital camera and it shows that the size of the homogeneous patterned area (the blue region, ~6 × 6 mm) can reach the same order of magnitude of the one of a coin of 2 cents of euro. The image reported in Fig. 6.5b is a zoom of the red dot in Fig. 6.5a and displays the pattern obtained by starting from a stamp with 1,500 nm pitch and 600 FWHM; the blue arrow indicates the direction of the grooves.

The homogeneity of the patterns over large areas (4×2 mm^2) has been assessed by the statistical analysis of different printed stripe arrays obtained with three different stamps as in Table 6.2.

The mean FWHM and height values of printed features were extracted from several AFM topographical images acquired with steps of 400 μm along the

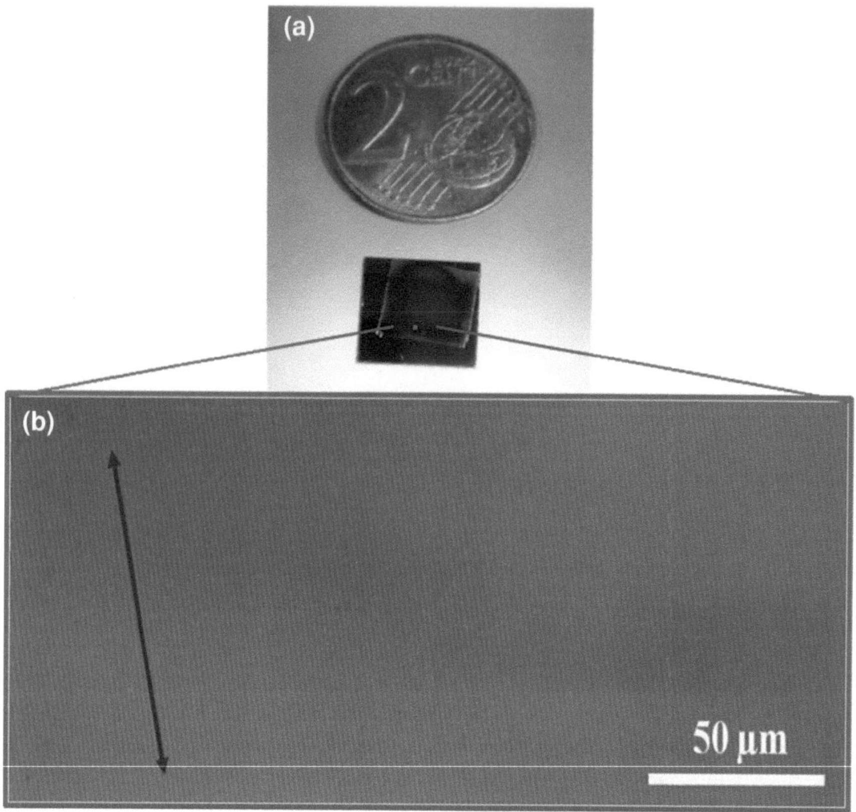

Fig. 6.5 Optical images displaying the patterned silicon oxide wafer next to a 2 euro cent coin (**a**), and a zoom on the linear pattern on the silicon oxide (**b**)

Table 6.2 Characteriztic feature sizes of PDMS stamp used for LCE	Stamp A	Stamp B	Stamp C
Pitch (nm)	1,500	710	520
FWHM (nm)	600	420	320

parallel and perpendicular directions with respect to the wires orientation (Fig. 6.6). The movement along each direction has been achieved by rotating micrometric screws beside the AFM head.

Figure 6.6a shows a representative "Histogram function" graph that can be extracted from a single AFM image (IAProject software, NT-MDT, Moscow, Russia). It represents a distribution of the heights of the topographical image; through this function, the mean height of the printed features in the image can be obtained. By iterating this analysis for each step and plotting the height values versus the scanning step, a graph as in Fig. 6.6b is obtained.

Although feature height results systematically slightly ($\sim$10%) higher at the borders, there is a large and predominant homogeneous area (>96%) where the mean

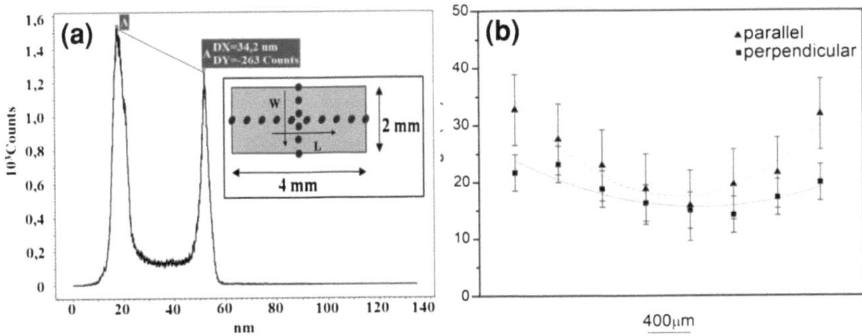

Fig. 6.6 **a** Example of "Histogram function" extracted from a single AFM image obtained by scanning the sample replicating stamp B, in according to the scheme reported in the inset. **b** Height of printed features measured along the direction perpendicular(▲) and parallel (■) to the stripes. Error bars represent the standard deviation

height (51.1 ± 3.4 nm starting from *stamp A* and 22.1 ± 2.2 nm from *stamp B*) does not change significantly ($p > 0.05$, One-Way ANOVA analysis, Tukey's post test, Prism5, GraphPad Software).

6.3 Simulation by Finite Elements of the Lithographically Controlled Etching Process

Simulation by finite element model of the etching process has been carried out to explore isotropic etching in confined environment [10]. The solution provides the flow dynamics within the channels employing the experimental etch rates reported in Ref. 8. In the simulation the stamp is in contact with the surface at the time $t = 0$. The rectangular (600×200 nm^2) channel section perpendicular to the direction of the wires is adopted in this simulation. Surface tension term has been included in the Navier–Stokes equation to calculate the evolution of the profile of silicon dioxide thin film in time (Fig. 6.7a). The reaction taking place on the surface generates gaseous SiF_4[11] which reduces the overall density of the fluid near the etched region approximately by 10%. A buoyancy effect allows the calculation of the fluid streamline field [12]. The isotropic pattern generally described in literature is now hindered by the submicrometric dimension of the channels and consequent HF dynamics therein. The concentration control the patterning sizes achieved within LCE process, therefore the concentration of HF has been estimated within the channels by convection and diffusion, using the velocity field obtained from Navier–Stokes equation. The initial volume of the etching solution is comparable with the volume occupied by the silicon dioxide thin film.

During the process the advancing front increases the volume of the cell, and HF solution dilutes (due to the chemical reaction of Eq. 6.1) creating a gradient of concentration and leading to slow down the etch rate (Fig. 6.7b). From the results,

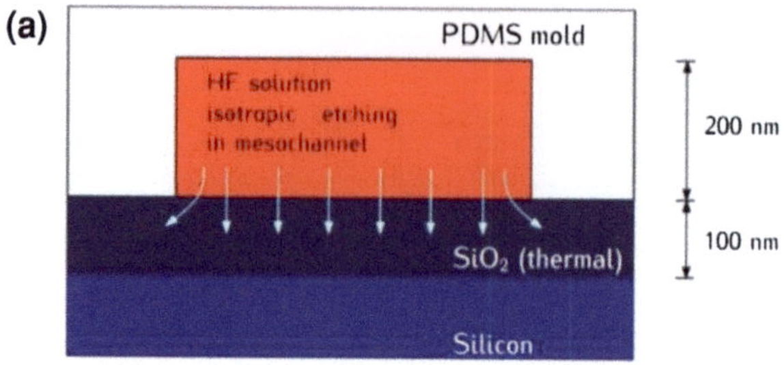

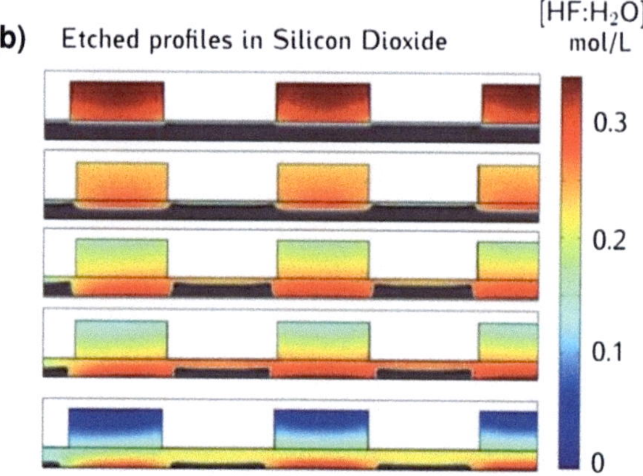

Fig. 6.7 LCE simulation by finite element. **a** Scheme of the subdomains used in finite element calculation. HF solution reacts with underlying silicon dioxide surface isotropically, although confined by the PDMS mold. **b** Time evolution of the concentration of the etching solution, within the mesoscopic channel section. At the beginning, the silicon dioxide thin film is etched in correspondence of the channel cavities, after the etching, the concentration is polarized towards the surface. The simulation has been carried out with 25 nm/min etching speed

it appears that at later stages there is a polarization of HF concentration toward the silicon surface due to the structuring of the substrate. Summarising, the simulation shows at the beginning of the process, the surface is preferentially etched in correspondence of the channel cavities and the etching solution concentration is homogeneous; after a few seconds, also the regions under the stamp protrusion are partially etched (Fig. 6.7b).

6.4 SiO$_2$ Nanowire Fabrication

The control of the etching time allows a fine tuning of the height and size of the submicrometric linear structures. By employing a PDMS stamp with narrower features (with respect to *stamp A* and *stamp B*, Table 6.2) and applying a further etching treatment in bulk, it is possible to fabricate silicon oxide array stripes less than 100 nm sized and millimiter long (Fig. 6.8).

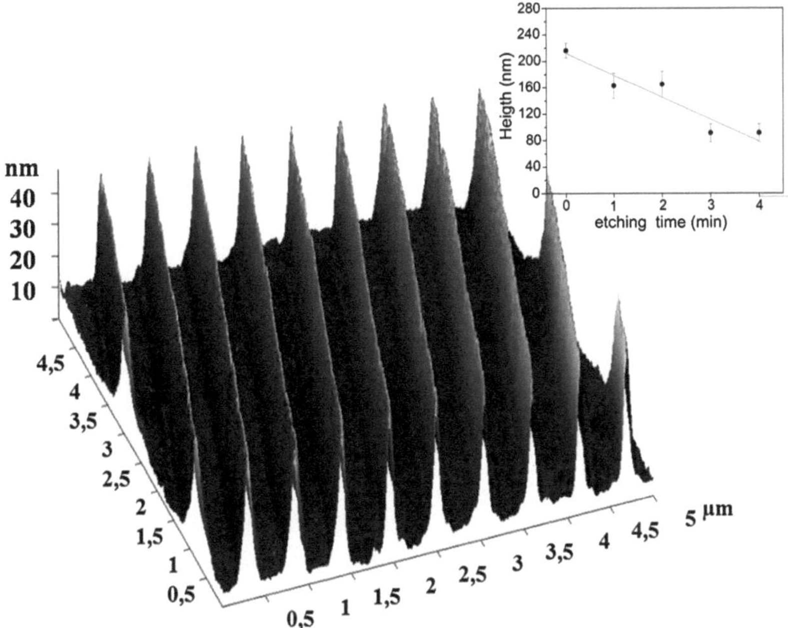

Fig. 6.8 3D AFM image of printed SiO$_2$ nanowires. The inset shows the width of printed nanowires versus etching time

Table 6.3 Characteriztic topographical properties of printed features

Sample	FWHM (nm)	Pitch (nm)	Stripe density (n × cm^{-2})	Duty cycle (%)	Aspect ratio
Sample 1	630	1,496	20,053	42.1	12.4 : 1
Sample 2	296	709	4,231	29.2	9.6 : 1
Sample 3	216	520	19,230	41.5	7.7 : 1
Sample 4	91	520	19,230	17.5	2.5 : 1

In Fig. 6.8 a bulk etching treatment (4% HF, 4′) has been applied to a pattern obtained by LCE using *stamp C* (Table 6.2) to produce a final pattern with stripes showing a mean FWHM of 91.3 ± 4.5 nm, i.e. SiO$_2$ nanowires.

Beside the 3D AFM image of the SiO$_2$ nanowires, a graph shows the evolution of stripe mean size versus etching time.

Characteriztic features (FWHM, pitch, stripe density, duty cycle, aspect ratio) of stripe arrays obtained by different PDMS stamps are summarized in Table 6.3.

The duty cycle is defined as the percentage ratio between FWHM and pitch. The aspect ratio in this case is defined as the ratio between FWHM and the height of printed stripes.

Sample 1 has been obtained from a *stamp A*, sample 2 from *stamp B*, sample 3 from *stamp C* and sample 4 (SiO$_2$ nanowires) from bulk etching after the first LCE treatment with *stamp C*.

Table 6.3 indicates that by LCE technique it is possible to fabricate linear patterns with high stripe density (20,000/cm^2) very low duty cycle (17.5%, with the printed stripe being about 6 time smaller than the pitch) and aspect ratio $\sim$2.5.

6.5 One-Pot Surface Patterning and Nanoparticle Deposition

An attractive feature of LCE is the possibility to deposit functional materials and/ or nanoparticles (NPs) at the same time of the etching process, simply by starting from a mixture of the ES and the functional materials. The main limitation is that the solute must be resistant to the ES.

The scheme of LCE combined with solute deposition is shown in Fig. 6.9.

Interestingly, LCE encompasses and integrates both top–down (as confined etching) and bottom–up (as self-assembly) nanofabrication approaches.

As can be noticed, LCE deposition reproduces similar conditions as unconventional wet lithographic techniques [13, 14], in particular it resembles the condition occurring in the last step of micromolding in capillaries [15]. As the critical concentration in the solution is reached, the solute precipitates inside the cell onto the substrate giving rise to a structured thin film replicating the stamp recesses. Using an enough concentrated solution, the latter reaches the supersaturation regime (the volume of the solution is comparable with the volume of the thin layer cell) and a homogeneous thin film of the dispersed materials is obtained. On the other hand, when using a diluted solution, the supersaturation regime is reached in a later stage of the evaporation, i.e. when the volume of the solution becomes smaller than the volume of the cell. In this regime the capillary forces pin the solution to form menisci around the protrusions. Figure 6.9a shows the deposition of colloidal un-conjugated gold nanoparticles (Au NPs, $\varnothing$ < 15 nm) in the pattern grooves, using a 4% HF solution mixed with 10 mg/L of water soluble Au NPs (with the acid and the nanoparticles in a 1:10 v/v ratio) using *stamp B*.

The amount of deposited NPs is controlled by adjusting the initial concentration of the solution. In Fig. 6.10a the NPs density is >10 NPs/μm^2 while in Fig. 6.10 b, c is <1 NPs/μm^2.

An interesting phenomenon of LCE combined with NPs deposition is shown in Fig. 6.10b and c. In this case the feature of the PDMS stamp are sub-micrometric logic structures formed by segments and dots (the same of a written DVD). Using a very diluted solution (corresponding to deposition of 30–50 NPs on 10 $\times$ 10 μm^2), the probability of finding a single nanoparticle close to a micrometric structure is proportional to the length of the structure. In fact the process is guided by the meniscus formation at the later stage of the evaporation that drives the solution around the protrusions in contact with the substrate [13]. In this way the use of a stamp containing a number of micrometric structures comparable with the number of NPs permits the positioning of a limited number of NPs only around the longer structures. From the analysis of about 100 structures it has been

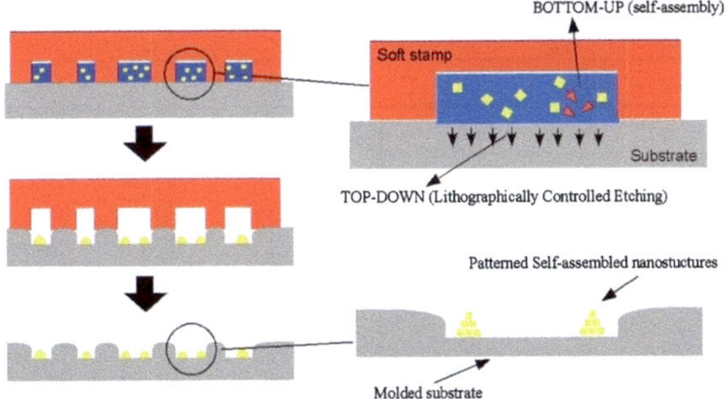

Fig. 6.9 Schematic illustration of LCE combined with functional material deposition

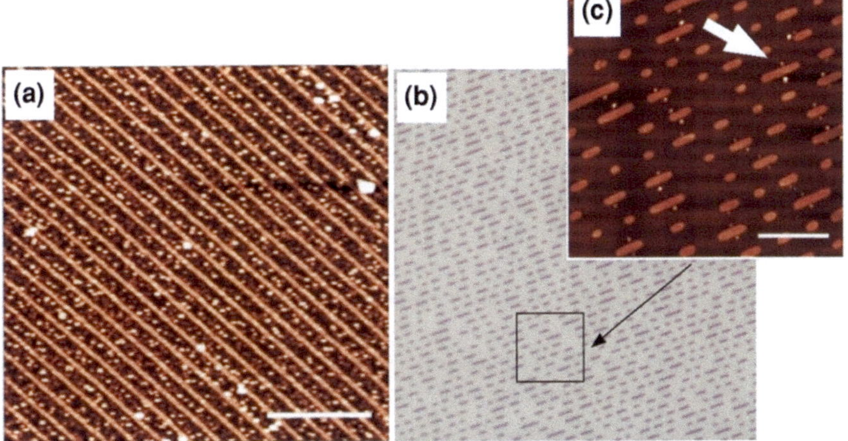

Fig. 6.10 LCE combined with Au NPs deposition. **a** AFM images of high density NPs deposited in the pattern grooves. Scale bar 4 μm. **b** Optical micrograph of low density NPs deposited next to logic structures. Scale bar 10 μm. **c** AFM image detail of **b**. The white arrow indicates NPs self-placed in correspondence of about the half of the longer (> 500 nm) structures

calculated that more than 98% of NPs are placed in correspondence of the middle position of the printed structures longer than 500 nm. Therefore, by using a simple optical microscope (Fig. 6.10b, a 50 X objective is more than enough), single NPs become addressable for further characterization or application, depending on their functionality.

The chemical composition of the surface, as well as the aggregation state and functionality preservation of Au NPs after processing has been assessed by X-ray Photoelectron Spectroscopy (XPS) and Absorption Spectroscopy.

Table 6.4 Atomic concentrations (%) and BE values (eV) of the chemical species on the surface of different representative samples

Peak→	C1	C2	C3	F 1s	O1s	Si1	Si2	Si3	N1s	Au4f7
BE, eV	285.0	≈ 287	≈ 289	688	≈ 533	≈ 103.4	≈ 105.4	≈ 102	≈ 400	84.0
Sample↓										
1	3.8	1.4	0.6	0.6	60.8	29.5	3.3	0	–	–
1a	20.5	7.2	0	0.4	51.6	20.0	0	0	0.2	0.085
2	24.6	2.9	0.8	0.4	40.1	14.7	2.6	13.9	–	–
3	7.7	0.5	0	0.8	56.8	34.1	0	0	–	0.1
4	19.9	1.5	1.0	–	47.2	24.7	0	5.3	0.1	0.2

6.6 X-Ray Photoelectron Spectroscopy

The atomic concentrations of constituent elements and the binding energy (BE) values of the main XPS peaks, including their synthetic components obtained by peak fitting, have been collected with an ESCALAB MkII (VG Scientific, UK) spectrometer and presented in Table 6.4.

Sample 1 and *sample 1a* are the reference samples: untreated SiO_2/Si wafer and the SiO_2/Si wafer with Au NPs (drop casting) respectively; *sample 2* is the LCE treated SiO_2/Si wafer; *samples 3* and *sample 4* are the LCE and LCW treated SiO_2/Si wafers with Au NPs, respectively. *Sample 4* has been introduced to assess the effect of HF on the chemical composition of the surface in the presence of Au NPS (*sample 3*), by comparing the latter with a sample prepared following a normal LCW deposition (thus using water rather than HF).

As can be inferred from Table 6.4, the surface of the untreated reference sample (*sample 1*) is characterized by a quasi-stoichiometric SiO_2 (Si1 and O peaks, 1 : 2) with a very low amount of contaminants (C and F). Only a small part of silicon (Si2) results to be bonded in some silicate and/or in some compound containing Si–C-F. Hence the initial cleaning protocol of the silicon wafer (see Sect. 6.8) has to be considered efficient. A similar composition has been detected for *sample 1a*, where the presence of deposited Au NPs is revealed along with contaminants (C and N) originating from the nanoparticle organic ligand (mainly sodium citrate), thus modifying the stoichiometry of the silicon oxide surface.

LCE treated wafers (*samples 2* and *3*) exhibit similar composition with mild presence of Si suboxides in the sample without Au NPs (*sample 2*). The smaller percentage of C contaminants in *sample 3* with respect to *sample 1a* can be ascribed to a partial elimination of the Au NPs organic ligand by HF. This hypothesis is confirmed also by data of *sample 4*, in which, in the absence of HF, the presence of C contaminant is significant.

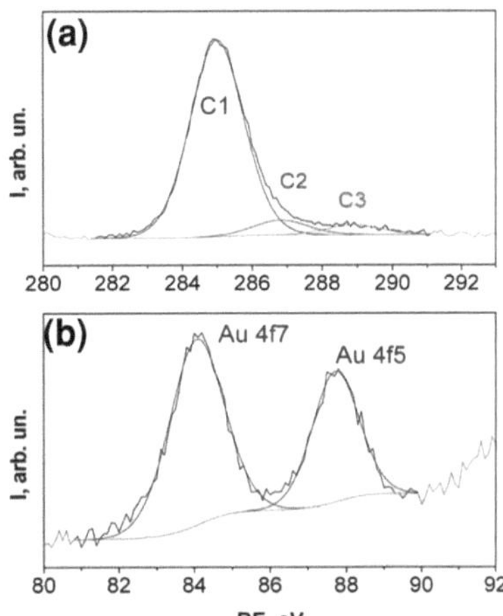

Fig. 6.11 XPS spectra of **a** C1 s, and **b** Au 4f regions of sample 3 prepared by LCE with Au NPs

Moreover, the low signal of N 1 s at BE $\approx$ 400 eV, registered in *samples 1a* and *4*, that indicates the presence of adsorbed amine groups[16] probably coming from the organic ligand, is not present in *sample 3*.

In Fig. 6.11 the characteriztic XPS peaks of *sample 3* are reported.

The main component C1 of C 1 s (Fig. 6.11a) is attributed to the most common C–C and C–H bonds, whereas the lower signals of C2 and C3 can be assigned to hydroxylic (CH–OH) and carboxylic (O = C–OH bonds), respectively [17]. The spectrum of Au 4f (Fig. 6.11b) is composed of typical spin–orbit splitted doublet. These peaks correspond to metallic gold Au^0 at BE = 84.0 and 87.7 eV. The positive Au species, that can be found in the case of NPs covered with a positively charged shell [18, 19], is completely absent in our samples. No sub-products, derived from the etching treatment, have been produced on SiO_2 substrates.

6.7 Absorption Spectroscopy

In order to assess the functionality of Au NPs upon LCE process, absorption spectra of processed NPs drop cast both from water solution and from 0,4% HF solution, have been recorded by a Lambda 950 spectrophotometer (Perkin–Elmer, USA). For this characterization 0.6 ml of Au NPs solution have been cast on quartz circular cylinders (diameter $\sim$1 cm, thickness $\sim$3 mm) through consecutive depositions, and dried at room temperature after each deposition.

Both absorption spectra (Fig. 6.12) exhibit the typical absorption peak of Au NPs at around 530 nm, which is characteriztic of isolated gold nanoparticles of 10–15 nm diameter [20], the same size of the the NPs used in these experiments. The peak of the spectrum relative to only Au NPs is more evident than the one

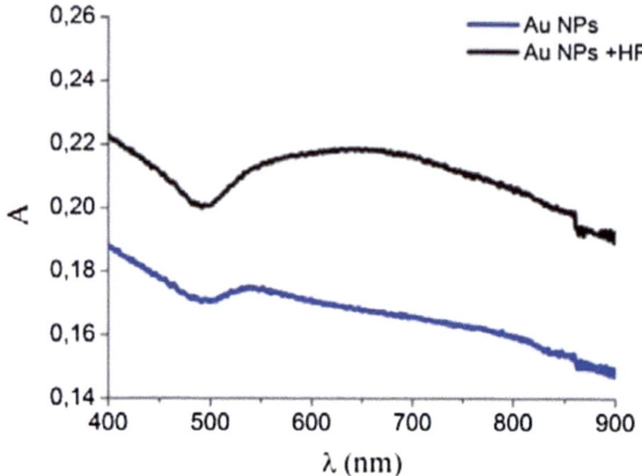

Fig. 6.12 Absorption spectra of Au NPs drop cast from water solution (*blue curve*) and processed in 0.4% HF (*black curve*)

with HF. In both cases the two spectra show a considerable spreading in the range 530–800 nm which indicated the presence of chains or aggregates of gold nanoparticles (expected using these concentration). It can be concluded that no relevant effect due to the process have produced, since a partial aggregation of NPs was already present in the starting material.

Finally, absorption spectroscopy measurements prove that LCE does not alter the macroscopic functionality of Au NPs.

6.8 Materials and Methods

6.8.1 Substrate, Stamps and Solutions

Substrates consist of a piece 8×8 mm^2 of silicon covered by 200 nm of thermal SiO$_2$ (Fondazione Bruno Kessler, Trento, Italy), have been cleaned by 2 min ultrasound in electronic-grade water (milli-pure quality), 2 min in acetone (Aldridch chromatography quality), then 2 min in 2-propanol (Aldridch spectroscopic grade quality). An air plasma treatment (Pelco easiGlow, USA) has been performed (t = 15 min, i = 30 mA) to further clean the surface eliminating all the organic compounds. Substrate resulted in a clean flat surface (RMS roughness = 0.25 ± 0.01 nm) with a water contact angle of $\theta_c = (61.8 \pm 1.5)°$.

Elastomeric polydimethylsiloxane (Sylgard 184 Down Corning, USA) stamps have been prepared by REM of commercials masters made of silicon (SCRIBA Nanotechnologie) and commercial Compact Disk (CD), Digital Versatile Disk (DVD) and Holographic Diffractive Grating (HPGs). The curing process has been carried out at 90°C for 3 h. After that the replica is peeled off from the master and cleaned with N_2. Logic structures have been prepared by replica molding of blank or written DVD.

Commercial SPI-Mark^{TM} unconjugated gold colloidal suspension (1 g/L, mean diameter <15 nm, SPI supplies Inc., USA) has been used to demonstrate the capability of the LCE technique to deposit material during the etching process.

The solutions have been prepared starting from HF (Sigma Aldrich 40% v/v) diluted in electronic-grade water (milli-pure quality). For the etching experiments without Au NPs, the final HF concentration was 0.4%. For the NPs deposition experiments, NPs have been diluted 1:10 in water and mixed with HF 4% (HF:NPs 1:10 v/v), thus the final HF concentration has been kept equal to the one used in the experiments without NPs.

6.8.2 Atomic Force Microscopy

Atomic Force Microscopy (AFM) images have been acquired with a NT-MDT Smena AFM (Moscow, Russia) operating in semi-contact mode under ambient conditions (relative humidity 55%). Single crystal silicon, n-type, 0.01–0.025 Ohm·cm, antimony doped NSG10 cantilevers with typical tip curvature radius of 10 nm have been used. All images are unfiltered, only a line-by-line background subtraction has been performed to remove trend effects by images treatment software (IAProject, NT-MDT; Moscow, Russia).

6.8.3 X-Ray Photoelectron Spectroscopy

XPS has been performed in order to assess the absence of etching residues on the surface and the presence of metal nanoparticles. These measurements were carried out by using an ESCALAB MkII (VG Scientific, UK) spectrometer equipped with a standard Al Kα excitation source and a 5-channel detection system. Photoelectron spectra were collected at 20 eV constant pass energy of the analyzer and a base pressure in analysis chamber of 10^{-8} Pa. The spectra were processed by the Casa XPS v. 2.2.84 software, using a peak-fitting routine with symmetrical Gaussian–Lorentzian functions. The background was subtracted from the photoelectron spectra by using Shirley method.

6.9 Conclusions

In conclusion a new soft lithographic fabrication technique named LCE, integrating top–down (as confined etching) and bottom–up (as self-assembly) methodologies, has been developed.

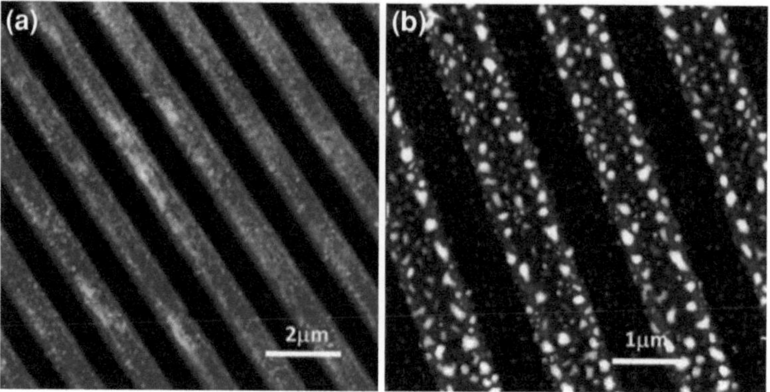

Fig. 6.13 **a** TiO$_2$ pattern by LCE starting from stamp A. **b** TiO$_2$ with magnetic nanoclusters embedded in the material

I have demonstrated the ability of LCE to fabricate micro and nanostructures, down to sub-100 nm nanowire arrays, over macroscopic areas, in a fast and reliable way.

I also proved that functional NPs can be deposited in addressable positions by a single step process, starting from a mixture of etching solution and NPS, thus overcoming the state of the art-fabrication techniques relatively to this problem. The possibility to deposit functional NPs within the cavities of the pattern in a controlled manner makes it suitable for biomedical sensing [21], optoelectronics and data storage applications.

6.10 Outlook of Lithographically Controlled Etching

Up to this extent, LCE technique has been applied to thermal SiO$_2$/Si wafer using HF as etching solution and gold nanoparticles as functional nano-object to be addressed.

Interestingly, LCE can be extended to other technological attractive materials like titanium dioxide (bone [22] and dental [23] implants, scaffolds for regenerative medicine [24], photocatalysis [25] and, photovoltaics [26]), and glass (substrate for cell growth [27] and biosensing [28, 29]).

In Fig. 6.13a the fabrication of a linear pattern of TiO$_2$ made by applying LCE to a thin TiO$_2$ film (thickness $\sim$150 nm) is displayed. The film has been fabricated by spin coating of a titanium dioxide precursor, named "bis(ammonium lactate)titanium dihydroxide" (TALH, 50% in water, Aldrich) followed by a thermal treatment at 450°C in air to get the anatase crystalline phase [30]. *Stamp A* (see Table 6.2) has been used in LCE to obtain the pattern in Fig. 6.12a. The patterning of a TiO$_2$ greatly enhances its superficial area, thus it could be interest for applications in which an increase of the active material area is desirable, as photovolatics and photocatalysis.

Fig. 6.14 Glass patterning
with stamp A. Bar is 2 µm

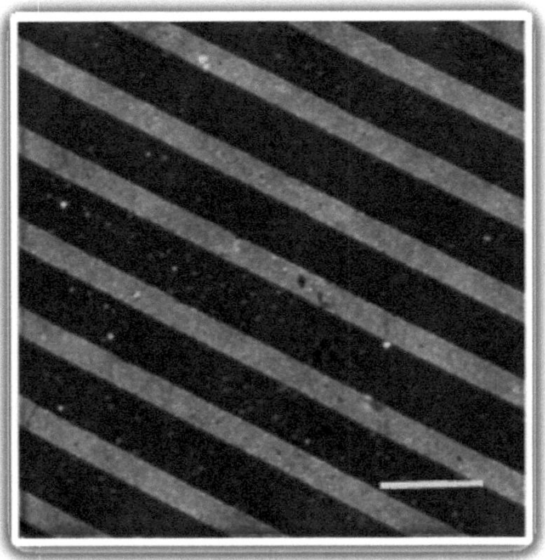

The AFM image in Fig. 6.13b shows a patterned TiO_2 thin film with ferromagnetic aggregates embedded. This pattern has been obtained following the same protocol as for Fig. 6.13a but this time starting from a mixture of TALH with a ferromagnetic spinel precursor $(\{[Co(H_2O)_2]_3[Cr(ox)_3]_2(18\text{-crown-}6)_2\}$ [31]. A pattern of this type can be thought to find possible applications in the field of spintronic devices or magnetic devices for data storage.

In Fig. 6.14 the patterning by LCE of a borosilicate cover glass slide (thickness ~1 mm, Thermo Fisher Scientific Inc., MA, USA) is displayed. The stamp used was *stamp A*.

The possibility to generate highly ordered pattern directly into glass surface, opens the way for a possible use of LCE in any application concerning cell growth, proliferation and differentiation study, as many supports for this scope are made of this material (petri dishes, glass slides for confocal microscope, optical guides for plasmon resonance experiments, etc.)

References

1. Cheng W, Park N, Walter MT, Hartman MR, Luo D (2008) Nat Nanotechnol 3:682–690
2. Kim H, Kim J, Yang H, Suh J, Kim T, Han B, Kim S, Kim DS, Pikhitsa PV, Choi M (2006) Nat Nanotechnol 1:117–121
3. Huang Y, Duan XF, Wei QQ, Lieber CM (2001) Science 291:630–633
4. Klauk H (2010) Chem Soc Rev 39:2643–2666
5. Cavallini M, Facchini M, Massi M, Biscarini F (2004) Synth Met 146:283–286
6. Bustillo JM, Howe RT, Muller RS (1998) Proc IEEE 86:1552–1574
7. Buhler J, Steiner FP, Baltes H (1997) J Micromech Microeng 7:R1–R13
8. Williams KR, Gupta K, Wasilik M (2003) IEEE J Microelectromech syst 12:761–778

9. Yahyaoui F, Dittrich T, Aggour M, Chazalviel JN, Ozanam F, Rappich J (2003) J Electrochem Soc 150:B205–B210
10. Son G (2010) J Mech Sci Technol 24:991–997
11. Knotter DM (2000) J Am Chem Soc 122:4345–4351
12. Greco P, Facchini M, Cavallini M, Ruani G, Dionigi C, Biscarini F (2010) Phys Status Solidi B 247:877–883
13. Cavallini M, Albonetti C, Biscarini F (2009) Adv Mater 21:1043–1053
14. Cavallini M, Facchini M, Albonetti C, Biscarini F (2008) Phys Chem Chem Phys 10:784–793
15. Kim E, Xia YN, Whitesides GM (1996) J Am Chem Soc 118:5722–5731
16. Sylvestre JP, Poulin S, Kabashin AV, Sacher E, Meunier M, Luong JHT (2004) J Phys Chem B 108:16864–16869
17. Svorcik V, Hubacek T, Slepicka P, Siegel J, Kolska Z, Blahova O, Mackova A, Hnatowicz V (2009) Carbon 47:1770–1778
18. Mazzaglia A, Scolaro LM, Mezzi A, Kaciulis S, De Caro T, Ingo GM, Padeletti G (2009) J Phys Chem C 113:12772–12777
19. Gillet JN, Meunier M (2005) J Phys Chem B 109:8733–8737
20. Duchesne L, Gentili D, Comes-Franchini M, Fernig DG (2008) Langmuir 24:13572–13580
21. Adamcik J, Viglasky V, Valle F, Antalik M, Podhradsky D, Dietler G (2002) Electrophoresis 23:3300–3309
22. Feng B, Weng J, Yang BC, Qu SX, Zhang XD (2003) Biomaterials 24:4663–4670
23. Gapski R, Wang HL, Mascarenhas P, Lang NP (2003) Clin Oral Implant Res 14:515–527
24. Wei J, Chen QZ, Stevens MM, Roether JA, Boccaccini AR (2008) Mater Sci Eng C-Biomimetic Supramol Syst 28:1–10
25. Carp O, Huisman CL, Reller A (2004) Prog Solid State Chem 32:33–177
26. Hagfeldt A, Gratzel M (2000) Acc Chem Res 33:269–277
27. Rhee SW, Taylor AM, Tu CH, Cribbs DH, Cotman CW, Jeon NL (2005) Lab Chip 5:102–107
28. Storhoff JJ, Lucas AD, Garimella V, Bao YP, Muller UR (2004) Nat Biotechnol 22:883–887
29. Thoumine O, Ott A (1997) J Cell Sci 110:2109–2116
30. Ruani G, Ancora C, Corticelli F, Dionigi C, Rossi C (2008) Sol Energy Mater Sol Cells 92:537–542
31. Coronado E, Marti-Gastaldo C, Galan-Mascaros JR, Cavallini M (2010) J Am Chem Soc 132:5456–5468

Chapter 7
Conclusions

In this thesis novel approaches for the fabrication of multiscale structure systems have been explored, with the aim of providing novel tools for the investigation of cellular behaviour in response to different external stimuli. In particular, the research activity has been focussed on the adhesion and proliferation of two neural cell lines (human SH-SY5Y neuroblastoma and 1321N1 astrocytoma) with respect to chemical, topographical and electrical stimuli. This is important for understanding the process of cell regeneration that is a key issue for central nervous system diseases and spinal chord injuries.

To investigate the cell response to chemical signals, patterns of a cell adhesive molecule such as laminin on completely non-adhesive surface like Teflon-AF have been fabricated, obtaining a cell adhesion selectivity close to 100%.

In order to assess the cell behaviour with respect to topographical features, a fabrication procedure based on biocompatible titanium dioxide patterned in the presence polystyrene nanoparticles as templating agent was applied. This approach accuratey regulates the topography/porosity of the pattern. This system revealed to be enough efficient to obtain information on the cell behaviour in response to local morphological alterations.

The very first steps of neural cell adhesion were studied in the presence of mild electric fields, to assess the role of the application of a current on the initial cell spreading over the surface. The system was realized by employing patterned single-walled carbon nanotubes, chosen for their attractive chemical-physical properties and allowed to appreciate a positive correlation between electric field value and cell spreading.

Finally a new patterning technique, named Lithographically Controlled Etching(LCE), has been proposed as suitable for micro/nanostructuring a technological surface with grooves and offering the possibility to simultaneously functionalise the grooves with molecules or nanoparticles. This happens within the same

M. Bianchi, *Multiscale Fabrication of Functional Materials for Regenerative Medicine*, Springer Theses, DOI: 10.1007/978-3-642-22881-0_7,
© Springer-Verlag Berlin Heidelberg 2011

process, thus eliminating one step with respect to the current lithographic protocols. LCE can be applied to different functional materials, and could be promising for applications in the fields of regenerative medicine and biosensing.